A PRACTICAL APPROACH TO SCIENCE AND ENGINEERING WITH SELF-REGULATED LEARNING

Science and engineering practices tend to be more difficult to teach and monitor than content knowledge, because practices are skill based. This book presents tangible ways for teacher educators and teachers to design learning environments that involve student goal setting, monitoring, and reflection on their performance of science and engineering practices. It models ways teachers can support effective learning behaviors and monitor student progress in science and engineering practices. It also presents practical ways to set up preservice teacher instruction and inservice teacher professional development that address both self-regulated learning and science and engineering practices. Educational research designs are presented from qualitative, quantitative, and mixed methods traditions that investigate student and teacher engagement with science and engineering practices through self-regulated learning.

ERIN E. PETERS-BURTON is Professor at George Mason University, USA, and Director of the Center for Social Equity through Science Education. She taught at secondary school for fifteen years prior to her academic work. She has published in science education, teacher education, educational psychology, marine biology, geology education, philosophy of science, technology, educational leadership, and learning disability journals. In 2016, she was named the ASTE Outstanding Science Teacher Educator of the Year.

A PRACTICAL APPROACH TO SUPPORTING SCIENCE AND ENGINEERING STUDENTS WITH SELF-REGULATED LEARNING

ERIN E. PETERS-BURTON
George Mason University

Shaftesbury Road, Cambridge CB2 8EA, United Kingdom

One Liberty Plaza, 20th Floor, New York, NY 10006, USA

477 Williamstown Road, Port Melbourne, VIC 3207, Australia

314–321, 3rd Floor, Plot 3, Splendor Forum, Jasola District Centre, New Delhi – 110025, India

103 Penang Road, #05-06/07, Visioncrest Commercial, Singapore 238467

Cambridge University Press is part of Cambridge University Press & Assessment, a department of the University of Cambridge.

We share the University's mission to contribute to society through the pursuit of education, learning and research at the highest international levels of excellence.

www.cambridge.org
Information on this title: www.cambridge.org/9781009100014

DOI: 10.1017/9781009103800

© Erin E. Peters-Burton 2024

This publication is in copyright. Subject to statutory exception and to the provisions of relevant collective licensing agreements, no reproduction of any part may take place without the written permission of Cambridge University Press & Assessment.

First published 2024

A catalogue record for this publication is available from the British Library.

Library of Congress Cataloging-in-Publication Data
NAMES: Peters-Burton, Erin E., author.
TITLE: A practical approach to supporting science and engineering students with self-regulated learning / Erin E. Peters-Burton, George Mason University.
DESCRIPTION: Cambridge, United Kingdom ; New York, NY : Cambridge University Press, 2023. | Includes bibliographical references and index.
IDENTIFIERS: LCCN 2023002771 (print) | LCCN 2023002772 (ebook) | ISBN 9781009100014 (hardback) | ISBN 9781009108270 (paperback) | ISBN 9781009103800 (epub)
SUBJECTS: LCSH: Science–Study and teaching (Secondary) | Engineering–Study and teaching (Secondary) | Self-managed learning. | Self-culture.
CLASSIFICATION: LCC Q181 .P3557 2023 (print) | LCC Q181 (ebook) | DDC 507.1/2–dc23/eng20230508
LC record available at https://lccn.loc.gov/2023002771
LC ebook record available at https://lccn.loc.gov/2023002772

ISBN 978-1-009-10001-4 Hardback
ISBN 978-1-009-10827-0 Paperback

Cambridge University Press & Assessment has no responsibility for the persistence or accuracy of URLs for external or third-party internet websites referred to in this publication and does not guarantee that any content on such websites is, or will remain, accurate or appropriate.

I would like to dedicate this book to my wonderful husband, Stephen, who not only supported me emotionally while I wrote, but had lots of conversations with me about science practices.

Contents

List of Figures	*page* ix
List of Tables	x
Acknowledgments	xi

PART I FOUNDATIONS

1	Student Engagement in Science and Engineering Practices	3
2	Unpacking Science and Engineering Practices	13
3	Self-Regulated Learning	28

PART II ENGAGING IN DISCIPLINARY TASKS IN SCIENCE AND ENGINEERING

4	Asking Questions and Defining Problems	47
5	Developing and Using Models	66
6	Planning and Carrying Out Investigations	88
7	Analyzing and Interpreting Data	110
8	Mathematics and Computational Thinking	131
9	Constructing Explanations and Designing Solutions	143
10	Engaging in Argument from Evidence	164
11	Evaluating and Communicating Information	185

PART III EDUCATIONAL RESEARCH AND TEACHER EDUCATION APPLICATIONS

12 Professional Development Designs — 209

13 Planning Lessons with Embedded Self-Regulated Learning Using the 5E Format — 228

14 Research Designs for Examining Science and Engineering Practices and SRL — 241

References — 255
Index — 262

Figures

1.1	Relationship of science and engineering practices with content and epistemic knowledge	*page* 9
2.1	Connections between engineering design processes and engineering practices	23
3.1	Phases and sub-processes in self-regulated learning theory	30
4.1	SRL processes for asking questions and defining problems	54
5.1	SRL processes for developing and using models	76
5.2	Diagram of electrolysis apparatus setup	79
6.1	SRL processes for planning and carrying out investigations	98
7.1	SRL processes for analyzing and interpreting data	119
8.1	Integration of computational thinking with data practices	133
8.2	SRL processes for mathematical and computational thinking	134
9.1	SRL processes for constructing explanations and designing solutions	152
10.1	SRL processes for engaging in arguments using evidence	174
11.1	SRL processes for evaluating and communicating information	195
12.1	Progression of instruction for scientific argumentation professional development	227
14.1	Maxwell's research design diagram	242
14.2	Research design diagram for case study	244
14.3	Research design diagram for quantitative comparison study	249
14.4	Research design diagram for mixed methods study	251

Tables

2.1	Relationships between science practices and nature of science aspects	*page* 17
2.2	Relationships between engineering practices and nature of engineering aspects	20
2.3	Task analysis table with an example activity	25
8.1	Abeer's data table	140
12.1	Topics, goals, and format for professional development on learning how to teach earth science through inquiry for grades K–5 by week	220
12.2	List of objectives and activities from a thirty-two-hour professional development program focused on learning argumentation and teaching argumentation in science	224
13.1	Connections between the 5E model of curriculum design and SRL phases	234
13.2	Task analysis of pendulums investigation	239

Acknowledgments

I would like to thank the Science Practices Innovation Notebook research team and teacher team. Without them I wouldn't have been able to have in-depth discussions about teaching science and engineering practices for the past five years. Thanks to Tim, Peter, Anastasia, Erin B., Erin W., Laura, Jake, Zach, Steph, Jessica, Connor, Hong, Britt, Suzanne, Matthew, Kat, Kevin, Lisa, Kim, Haley, Angela, Melissa, Charmaine, Emily, Candace, Swapna, Katie, and Jin. I am deeply grateful for being able to work with you all. Thanks, too, to my farm animals: Piobar, Chester, Mallow, Gusty, Noisy, Hank, Fia, Gabhy, Stormy, Bridget, and the chickens for spending my breaks between chapters with me.

PART I

Foundations

CHAPTER I

Student Engagement in Science and Engineering Practices

During my career as a secondary science teacher, teacher educator, and teacher education researcher, I set a primary goal of helping my students to learn how to learn. If they could become independent learners in class, then not only could they become efficient and effective learners in my science and science education courses, they could also transfer these skills to other topics. There is something powerful in being able to control your own learning, and it can open pathways to new skills and content knowledge that you never knew existed.

In order to help my students learn how to learn, I integrated into the disciplines of science and engineering a learning theory from educational psychology called self-regulated learning (SRL; Zimmerman, 2000). Self-regulated learning is a systematic method that looks at the way one learns, and the theory explains tangible processes that a learner uses to optimize their strategies. It has been shown over the years to be a very flexible theory, and has been used in many different subject matters and contexts such as writing, sports, science, and mathematics (Corno & Mandinach, 1983; Rohrkemper, 1989; Ryan, Connell, & Deci, 1984; Wang & Peverly, 1986; Zimmerman & Kitsantas, 2002). Self-regulated learning strategies have been taught to students to help them learn factual content knowledge, but there are other areas in which to support science and engineering students, namely, disciplinary approaches used while pursuing science and engineering (Duschl & Bybee, 2014; Pleasants & Olson, 2019).

Information about content knowledge of a subject is relatively easy to find on various platforms. However, defining and applying practices is more difficult. Understanding the role of practices and how to perform practices in the discipline allows a learner to gain a deeper level of knowledge, because a person can do science and engineering if they understand the practices. This gives them the power to find out relationships between variables on their own. Student understanding of science and engineering practices also goes hand in hand with self-regulated learning. Students who understand how to go about asking questions, developing procedures, gathering evidence, and communicating solutions can pursue problem-solving independently. The addition of SRL skills to the ability to perform science and engineering practices can result in students who can self-motivate, set productive goals, monitor their progress, and reflect on productive and unproductive processes. In effect, actively overlaying SRL onto learning about science and engineering practices can amplify the accuracy and efficiency of how students problem-solve while implementing disciplinary approaches.

The purpose of this book is to help teachers, teacher educators, and teacher education researchers establish and execute learning environments that support science and engineering students as self-directed learners. Teachers can use the ideas in the book to model and support student SRL, as well as to design explicit and reflective classrooms for students learning science and engineering practices. Teacher educators can use the ideas in the book to teach preservice teachers how to design learning environments that model, support, and assess student SRL and knowledge about science and engineering practices. Teacher education researchers can use the book to design research methodologies to investigate how teachers and students go about using SRL to learn science and engineering practices. The book directly addresses the teaching of primary and secondary students (aged 5–18) because that is what I have experienced in my career. However, the ideas in the book can be adjusted developmentally for undergraduate learners.

1.1 What Makes a Scientist a Scientist?

In order to self-regulate their learning, a student needs to have a context of learning about something, some skills or knowledge. Since this book is focused on science and engineering learning, it is important to know how science is defined as a discipline. In other words, what makes science a field of study? Science is treated separately from engineering here because they have different aims. A major aim of science is to explain phenomena that occur in the natural world. This is different from one of the major aims of engineering, to support human needs, which will be discussed in more detail in Section 1.2.

There has been a great deal of research on what makes science a unique discipline, often called the nature of science (NOS; Osborne et al., 2003). Within this research there are arguments about the level of detail to pursue in order to describe science as a discipline. For example, is science defined enough as a general discipline or should we be looking deeply into content areas such as biology, chemistry, and geology? An examination of standards of learning regarding NOS for all fifty states in the United States revealed that K-12 schools in the United States are treating science as a single, general discipline which is represented by overlapping aspects of NOS from various educational research frameworks (McComas, 2019). An applied approach to teaching NOS in K-12 classrooms would focus on the following overlapping aspects found in the standards:

- Science uses empirical evidence to make claims
- Scientific knowledge is mostly stable but tentative when new theories, models, and evidence are agreed upon
- History and societal norms influence knowledge production in science
- Science and technology have different aims, but support each other's development
- Scientists use creativity, critical reasoning, curiosity, and healthy skepticism in investigations
- Scientists work collaboratively and have professional standards that include ethical standards
- Scientific knowledge requires peer review

Although these aspects are important for all K-12 students to understand so that they can evaluate scientific claims and comprehend what is valued in scientific endeavors, the aspects tend to be philosophically oriented and may not be helpful in guiding students during investigations in a practical way.

Quality K-12 science instruction strives to mimic the ways scientists go about their investigations, but there are distinctions between the science students do in school and the science that professional scientists engage in (National Research Council, 1996). For example, professional scientists have a great deal of content knowledge and focus on specializations, whereas K-12 science students are generalists and are often learning each grade level's particular science knowledge for the first time. Although science students have life experience, they do not have the background knowledge that professional scientists possess. Professional scientists know their specialized field well and investigate questions about things that are currently murky, unknown, or on the fringes of the knowledge base. Students, on the other hand, usually investigate ideas that are accepted by the scientific community in order to understand them more thoroughly and build a base of knowledge in the field. Professional scientists differ from science students in that they share information globally by attending conferences and working in collaborative groups. Student scientists may be working in groups, but it is often for the purpose of developing communication skills, collaboration skills, and science content knowledge.

There are distinctions between professional science and school science, but there are also common aspects to both domains. The commonalities lie in science practices, approaches to investigations, or inquiries that exemplify habits of mind and methods that are valued by the discipline of science. As categorized by the Next Generation Science Standards (NGSS Lead States, 2013) in the United States, science practices are as follows:

- Asking questions
- Developing and using models
- Planning and carrying out investigations
- Analyzing and interpreting data
- Using mathematics and computational thinking
- Constructing explanations
- Engaging in argument from evidence
- Obtaining, evaluating, and communicating information

Students may not have as much background knowledge or expertise as professional scientists but they can still perform science practices, such as asking scientific questions, developing and using models based on their growing baseline knowledge, and engaging in argument from evidence.

1.2 What Makes an Engineer an Engineer?

As in the field of science education, engineering educators have explained what makes engineering a unique discipline. Based on the Accreditation Board for Engineering and Technology (ABET, 2021) and the National Academy of Engineering and National Research Council report, *Engineering in K-12 Education: Understanding the Status and Improving the Prospects* (2009), the nature of engineering can be described as what engineers do in the cyclical design process, how engineering impacts society, and how society impacts engineering. Topics in K-12 engineering education tend to emphasize engineering design, incorporating mathematics, science, and technology knowledge and skills, and promoting engineering habits of mind. Pleasants and Olson (2019) conducted a review of literature to develop a conceptual framework for the nature of engineering and found these disciplinary features:

- Design in engineering
- Specifications, constraints, and goals

- Sources of engineering knowledge
- Knowledge production in engineering
- The scope of engineering
- Models of design processes
- Cultural embeddedness of engineering
- The internal culture of engineering
- Engineering and science

Like NOS, the nature of engineering has strong conceptual foundations, but in order to be able to help students self-regulate their learning, the learning tasks need to be practical and observable. Self-regulated learning requires a learner to be able to set tangible goals, monitor those goals, and reflect on the outcome in relation to the goals. Again, a practical solution to teaching students to think like engineers is to focus on engineering practices. The Next Generation Science Standards have incorporated engineering practices into the standards that overlap with the science standards:

- Defining problems
- Developing and using models
- Planning and carrying out investigations
- Analyzing and interpreting data
- Using mathematics and computational thinking
- Designing solutions
- Engaging in argument from evidence
- Obtaining, evaluating, and communicating information

Science and engineering practices both depend on iterative cycles of inquiry that are governed by rational and logical thinking that lead to valid information. Science practices can guide students to understanding the natural world. Engineering practices can guide students to solving human needs. When students master science and engineering practices, they have a framework for problem-solving in many different contexts and develop the ability to refine their skills through conducting investigations.

1.3 Content, Procedural, and Epistemic Knowledge

As described earlier, professional scientists and engineers possess a great deal of background knowledge. This background knowledge is not only about content, but it is also about methods, practices, and rationales. Science and engineering practices are difficult to learn in a vacuum and require some content knowledge as a foundation, as well as knowledge of the rationale for using a practice in a particular situation. These three types of knowledge are known as content knowledge, procedural knowledge, and epistemic knowledge. Content knowledge includes the body of factual knowledge known as a discipline (e.g. science, mathematics, and engineering). Procedural knowledge is the understanding of how something is accomplished (e.g. science and engineering practices). Epistemic knowledge is understanding how an expert in a discipline thinks and what is valued in the discipline (e.g. NOS to nature of science). All three types of knowledge are important to learn so that one has flexibility and expertise in that type of disciplinary thinking (Osborne, 2014). A learner could learn about content knowledge alone, but it amounts to a collection of trivial facts. A learner could learn about procedural knowledge alone, but without content knowledge, it would be like steps of a process that have no goal. Learning about content and procedural knowledge without epistemic knowledge can result in a learner robotically performing practices to learn content but with no disciplinary guidance. All three types of knowledge are required to truly understand a discipline.

This book is structured so that teachers can address all three types of knowledge. Teacher educators and teacher education researchers can use the book to help teachers structure learning environments so that students understand how to think like a scientist and an engineer (epistemic knowledge) through understanding how science and engineering practices are performed (procedural knowledge). Content knowledge is addressed to a lesser extent, but is still present in the application of design challenges and investigations throughout the second part of the book.

1.4 Practices as a Lynchpin for Connecting Content and Epistemic Knowledge

Content, procedural, and epistemic knowledge all have their role in developing well-rounded learners. Self-regulated learning has shown to be helpful in supporting student learning of content knowledge (DiBenedetto & Zimmerman, 2010; Peters, 2012), epistemic knowledge (Peters & Kitsantas, 2010), and procedural knowledge in the form of science and engineering practices (NGSS Lead States, 2013). The focus of this book is on science and engineering practices because they can be the lynchpin for connecting content, epistemic, and procedural knowledge.

As demonstrated in Figure 1.1, there is a bi-directional relationship between content knowledge and science and engineering practices, and between epistemic knowledge and science and engineering practices. Students ask questions, set up investigations, analyze data, and communicate results (procedural knowledge) that can elaborate their current content knowledge. Conversely, prior content knowledge about the phenomena being investigated helps a science and engineering student effectively focus their use of practices. The outcomes of the practices can demonstrate disciplinary knowledge of science and engineering, particularly if the student is explicitly monitoring the ways the practices are being followed (e.g. systematically). Epistemic ideas can also help students direct their practices toward a particular goal, such as extending knowledge or solving a problem for a human need.

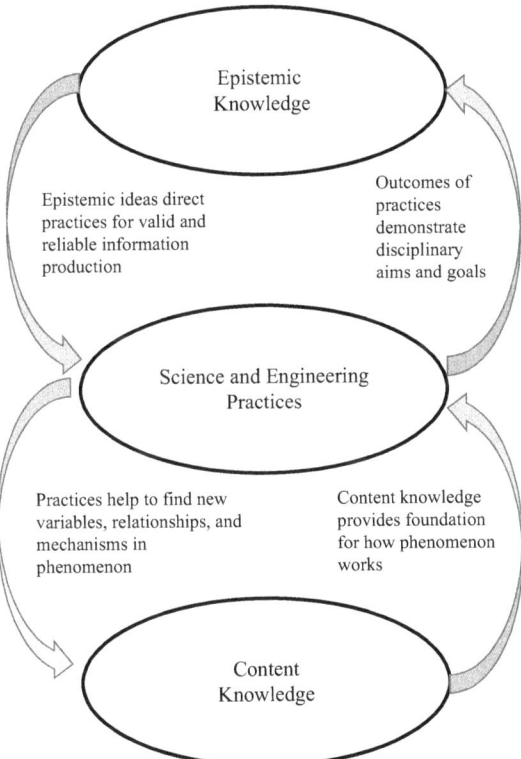

Figure 1.1   Relationship of science and engineering practices with content and epistemic knowledge

Consider the following scenario as an example. Celia is a student who understands the functionality of simple circuits and who is proficient in science and engineering practices. She is beginning an investigation that examines the question: How are parallel circuits different from simple circuits? Celia's content knowledge about simple circuits consists of the following facts:

- Simple circuits have a power source, a resistor such as a lamp, and one loop of conducting material
- The power must be strong enough and the loop must be complete and closed for the lamp to light

Celia is presented with a parallel circuit that has two branches, one power source and two lamps. She applies her content knowledge about simple circuits and notices that there is more than one loop in the parallel circuit. She can then focus her procedure design for the investigation on how the circuit behaves when each lamp is taken out of the circuit. Her content knowledge provides the foundation for how the phenomena might work and which variables she can manipulate to help her answer the research question. Once she tries to take out one lamp, her epistemic understanding that all conditions should be tested helps her to create more complete procedures. Once she has completed the investigation, she can reflect on how she used valid procedures in her investigation that can refine her epistemic knowledge. Celia's content, procedural, and epistemic knowledge allowed her to look at the investigation from multiple perspectives. The addition of explicit self-regulated learning processes could help develop her knowledge in a more systematic way.

1.5 Relationship of Self-Regulated Learning to Science and Engineering Practices

Self-regulated learning is inherently a problem-solving process. As explained in Chapter 3, the processes in the cyclical phases of SRL are essentially variables in a learner's toolkit, which can be measured and manipulated to change learning outcomes. SRL has three phases that occur (a) when a learner prepares for the learning task, (b) when a learner performs a learning task, and (c) when a learner evaluates the outcome of learning and adapts as needed. A cycle of SRL has parallels to both scientific inquiry and the engineering design processes. Chapter 3 goes into further detail regarding the processes that are associated in each phase of SRL and how they parallel different science and engineering practices.

When a student is more aware of their learning processes, they can treat the process of learning as a science investigation or engineering problem, thus reinforcing their understanding of science and engineering practices while improving their learning processes.

1.6 Structure of the Book

The book is structured into three main parts: (a) *Foundations*, (b) *Engaging in Disciplinary Tasks in Science and Engineering*, and (c) *Educational Research and Teacher Education Applications*. This book is written so that you can treat the book as a user manual rather than needing to read the book cover to cover, although that is also an option. If a reader is interested in supporting students on a particular practice such as asking questions and designing solutions with self-regulated learning, they can read Chapter 4 for background and strategies for that particular practice. However, if the reader is not familiar with self-regulated learning, they may want to read Chapter 3 before proceeding to the chapters on application of SRL theory to the practices. Likewise, if a reader is unfamiliar with the practices, they may want to read Chapter 2 before moving on.

The *Foundations* part includes chapters that discuss background ideas in science and engineering practices, self-regulated learning, and explains the overlap between the two realms. The *Engaging in Disciplinary Tasks in Science and Engineering* part has chapters that are dedicated to each science and engineering practice. In this part of the book, the chapters are organized by the practices found in the Next Generation Science Standards. Each of the chapters analyze the practice and articulate the skills that comprise the practices beyond what is available in the standards documents. From the detailed analysis of the practices, the chapters explain a way to help students self-regulate their knowledge of the skills that make up the practice, provide a positive and a negative case study of the practice, and offer questions for teachers to consider in order to adapt ideas for their classroom.

The *Educational Research and Teacher Education Applications* part of the book provides ideas for professional development designs based on a review of the research for preservice and inservice teachers in elementary and secondary settings. This part also describes example lessons using the 5E lesson planning format that embeds self-regulated learning. Finally, this part offers qualitative, quantitative, and mixed-methods research designs for studying student engagement in science and engineering practices supported by self-regulated learning.

This book is designed for science and engineering teachers, teacher educators, and educational researchers. Science and engineering teachers can use the *Foundations* and *Engaging in Disciplinary Tasks in Science and Engineering* parts to set up learning environments that support student engagement in science and engineering practices. Teachers can also use the *Educational Research and Teacher Education Applications* part to help design action research projects that can answer questions about student learning in their classrooms. Teacher educators can use the *Foundations* and *Engaging in Disciplinary Tasks in Science and Engineering* parts for instruction with preservice teachers and professional development experiences with inservice teachers. Teacher educators can use the *Educational Research and Teacher Education Applications* parts to design research to assess their instruction and to add to the teacher education literature. Educational researchers can learn more about the phenomena they investigate in the *Foundations* and *Engaging in Disciplinary Tasks in Science and Engineering* parts and use the *Educational Research and Teacher Education Applications* parts to design their own research for the field.

CHAPTER 2

Unpacking Science and Engineering Practices

The purpose of this chapter is to examine science and engineering practices in detail. The analysis of the practices aims to draw up connections with the practices' disciplinary characteristics and decomposes the learning tasks that can be accomplished to master science and engineering practices. A better understanding of how science and engineering practices represent (or do not represent) disciplinary characteristics elevates the practices beyond steps students do to get an "answer" for their investigation. Similarly, this chapter will examine practices through the lens of process and outcome goals so that teachers can use the decomposed learning tasks in each science and engineering practice to model disciplinary work to support students. Students who have little exposure to science and engineering disciplines are often at a disadvantage when asked to conduct an investigation because they may just follow the steps of a newly learned practice without knowing why they are doing the practices. To help students get beyond going through the motions, teachers can model science and engineering practices while pointing out why they are doing the practice in a particular way and why they are using the practice in the context. By breaking down practices, teachers can help students set smaller, more short-term goals that can support mastery of science and engineering practices.

It is unreasonable to expect a teacher to teach all eight science and engineering practices for each lesson, particularly when you are introducing them for the first time. It is likely that there will not be any lessons that feature all of the science and engineering practices. Rather, the expectation is that a teacher would examine a lesson and find one or more science and engineering practices to enhance in the context of the lesson. There are certain lessons that lend themselves to a particular science and engineering practice more effectively than others. Additionally, the practices are not intended to be taught in lockstep, beginning with asking questions and defining problems and ending with evaluating and communicating information. Some educational research indicates that the practice of developing and carrying out investigations may be the most accessible practice for some students (Duschl & Bybee, 2014). However, there is no one-size-fits-all with teaching practices. Teachers may want to begin with analyzing data that has already been collected, add on argumentation, and then move to asking questions, designing procedures, and analyzing data. Mathematical and computational thinking may be woven throughout. This book is intended to serve as background information that can be tailored to fit the different contexts experienced by teachers and teacher educators.

2.1 Science and Engineering Practices

The *Next Generation Science Standards* (NGSS; NGSS Lead States, 2013) has identified eight science and engineering practices that are applicable for students from kindergarten to twelfth grade (aged 5–18). The practices are listed with a description below. According to NGSS, science practices can be distinguished from engineering practices by asking a question about the purpose of the practice. If a learner is attempting to answer a question about the natural world, then the practice is likely science. If a learner is attempting to solve a problem that serves a human need, then the practice is likely engineering.

2.1.1 *Asking Questions and Defining Problems*

Students ask questions about the natural world that can be addressed with empirical evidence. Students define problems by understanding the scope of the problem, recognizing constraints and resources.

2.1.2 *Developing and Using Models*

Students develop and refine models as they learn to understand a system or parts of a system. Models are representations of the phenomenon of interest.

2.1.3 *Planning and Carrying Out Investigations*

Students set goals and develop and implement a plan to reach the goal of describing a phenomenon, testing a model, or developing an elegant solution.

2.1.4 Analyzing and Interpreting Data

After conducting the investigation and obtaining empirical data, students find patterns and make meaning from the data related to the question or problem being pursued.

2.1.5 Using Mathematics and Computational Thinking

Students use mathematics to describe models, carry out investigations, and interpret data. Students use computational thinking such as decomposition, pattern finding, abstraction, algorithmic thinking, and automation in the investigation.

2.1.6 Constructing Explanations and Designing Solutions

Students make connections across the practices to make meaning of their findings in relation to the natural phenomenon or problem.

2.1.7 Engaging in Argumentation from Evidence

Students use argumentation as a process to solve problems or create explanations. Arguments include making a claim, backed up by evidence and reasoning.

2.1.8 Obtaining, Evaluating, and Communicating Information

Students are prudent consumers of technical and scientific texts. Students can also communicate scientific and technical information in a variety of ways.

As discussed in Appendix F of the *Next Generation Science Standards*, students should engage in practices at every grade level in a developmentally appropriate way. A summary of the breakdown of developmental practices by grade bands is found at the National Science Teaching Association's website for NGSS https://static.nsta.org/ngss/MatrixOfScienceAndEngineeringPractices.pdf. Examples of developmentally appropriate practices can be found throughout Chapters 4 through 11 in the learning task breakdown for each practice and in the sample lesson on pendulums found in Chapter 13.

2.2 How Can Science and Engineering Practices Help Students Understand Disciplinary Characteristics?

In Chapter 1, I made an argument that understanding epistemic knowledge from a discipline was important, but could not stand alone to develop deep knowledge. Epistemic knowledge needs to be linked to knowledge such as conceptual and procedural knowledge for a student to be a flexible thinker. The purpose of this section is to illustrate connections between the nature of science (epistemic) and the science practices (procedural) and the nature of engineering (epistemic) and engineering practices (procedural). I believe that explicitly making these connections can help educators be purposeful about designing instruction with science and engineering practices.

The intention of teaching science and engineering practices is to have students experience them in every grade level, and if teachers are not purposefully selecting them, certain practices may be overlooked. In addition, educational research has demonstrated that explicitly and reflectively teaching the nature of science and the nature of engineering is more effective than expecting students to implicitly learn epistemic knowledge while conducting investigations (Akerson, Buzzelli, & Donnelly, 2017; Khishfe & Abd-El-Khalick, 2002; Peters-Burton, 2015; Peters & Kitsantas, 2010). Mapping science and engineering practices to disciplinary aspects can assist educators with instructional design. Because the nature of science and nature of engineering aspects help to explain why a particular practice is done, teachers can help students understand the bigger picture and illustrate how scientists and engineers think during an investigation.

Table 2.1 displays the science practices that are related to the aspects of the nature of science described in Chapter 1. When students begin an investigation by asking questions, a teacher can point out how questions that are asked in science have changed over time to demonstrate how history and societal norms influence knowledge production in science. Teachers can also point out that students need to be creative and curious in order to ask unique questions, but still use critical reasoning to keep the questions within the boundaries of what science can answer. Students asking scientific questions should also possess healthy skepticism so that they keep bias in check as they proceed through investigations.

Teachers can support student learning with the use of conceptual models. For example, teachers can ask students to draw out a conceptual model of how the phenomenon of interest works at the beginning of an

Table 2.1 *Relationships between science practices and nature of science aspects*

Science practice	Connection to nature of science aspect
Asking questions	• History and societal norms influence knowledge production in science • Scientists use creativity, critical reasoning, curiosity, and healthy skepticism in investigations
Developing and using models	• Scientific knowledge is mostly stable but tentative when new theories, models, and evidence are agreed upon • Scientists use creativity, critical reasoning, curiosity, and healthy skepticism in investigations • Scientific knowledge requires peer review
Planning and carrying out investigations	• Science uses empirical evidence to make claims • Science and technology have different aims, but support each other's development • Scientists use creativity, critical reasoning, curiosity, and healthy skepticism in investigations • Scientists work collaboratively and have professional standards that include ethical standards
Analyzing and interpreting data	• Science uses empirical evidence to make claims • Scientists work collaboratively and have professional standards that include ethical standards
Using mathematics and computational thinking	• Science uses empirical evidence to make claims
Constructing explanations	• Science uses empirical evidence to make claims • Scientific knowledge is mostly stable but tentative when new theories, models, and evidence are agreed upon • History and societal norms influence knowledge production in science • Scientists use creativity, critical reasoning, curiosity, and healthy skepticism in investigations • Scientific knowledge requires peer review
Engaging in argument from evidence	• Science uses empirical evidence to make claims • Scientific knowledge is mostly stable but tentative when new theories, models, and evidence are agreed upon • History and societal norms influence knowledge production in science • Scientific knowledge requires peer review
Obtaining, evaluating, and communicating information	• Science uses empirical evidence to make claims • Scientific knowledge is mostly stable but tentative when new theories, models, and evidence are agreed upon

Table 2.1 (*cont.*)

Science practice	Connection to nature of science aspect
	• History and societal norms influence knowledge production in science • Scientists work collaboratively and have professional standards that include ethical standards • Scientific knowledge requires peer review

investigation, and then ask students to update the model as they gather new information from the investigation. As students update their models by changing or adding features to the phenomenon, teachers can point out that a feature of science is that the knowledge is mostly stable, but when new information is learned in the scientific community, the accepted models and theories can change. Developing models, like asking questions, requires creativity and curiosity, while making sure the model is logical and students use rational thinking so that they do not accept ideas without empirical backing. Teachers can also point out that when students explain and share their model with others, they are asking peers to review their work, which is highly valued and necessary for scientific knowledge production.

When planning and carrying out investigations, teachers can point out to students that the reason for conducting the investigation is to gather evidence, which is necessary to make claims about how the natural world works. Again, in designing procedures, teachers can explicitly point out that students are using creativity, critical reasoning, and curiosity in designing their work. If students are using technology to measure the phenomenon, teachers can ask how the technology is helping to further science and what the scientific principles are that govern the technology. Teachers can also help students understand that scientists work collaboratively and have professional standards that include ethical benchmarks by discussing with students how they are not being influenced by their prior expectations for findings when designing the procedures of an investigation. Also, teachers can set up expectations for roles when students work collaboratively so that each person in the working group is interdependent with others.

Similarly, when students analyze and interpret data from their investigation, teachers can point out that when making claims, students must closely connect their ideas with the empirical evidence they found from their investigation. Using mathematical and computational thinking in interpreting the evidence helps them to organize or break down the information in ways that are familiar to the discipline. Teachers can support students by reminding them to think like scientists by coming to consensus with others regarding their analysis by relying on the evidence and ethical interpretation of the evidence. Students should not ignore evidence because it is convenient or because it does not agree with students' initial beliefs about the phenomenon. Anomalous data is a teachable moment for students because there may be some issue with the design or implementation of the procedure that could logically lead to unexpected outcomes.

When students are constructing explanations or creating scientific arguments from the findings of their investigation, teachers can help students understand disciplinary thinking by pointing out that although they are using creativity and exercising curiosity their explanations and claims must be closely tied to the empirical evidence that came from the investigation. The empirical evidence that comes from the investigation is most valuable when the procedures are designed to produce valid and reliable data. Teachers can also ask students to peer review each other's explanations and claims, emphasizing that students have a healthy skepticism toward explanations and claims and to ask questions about the evidence when needed. Teachers can also point out the connections of the nature of science with practices when they ask students to think about scientific claims historically and how their findings might fit with the historical knowledge of the phenomenon they investigated.

Table 2.2 displays connections of the engineering practices to the nature of engineering aspects. Engineering teachers can help students understand the nature of engineering aspects by explicitly pointing out how they relate to engagement in engineering practices. When students are defining problems for an investigation, teachers can explain how scoping a problem consists of clearly defining specifications, constraints, and goals in the discipline of engineering, which is also an aspect of the internal culture of engineering. The problem that students are defining is typically for a human need, and cultural norms influence whom the need will serve and how it might solve problems for some people but create problems for others. Defining a problem also requires some working knowledge of science. Physics, materials science, chemistry, and sometimes biological science has influence on the way the problem has occurred and how the solution will work to address the problem.

Table 2.2 *Relationships between engineering practices and nature of engineering aspects*

Engineering practice	Connection to nature of engineering aspect
Defining problems	• Design in Engineering • Specifications, Constraints, and Goals • The Scope of Engineering • Cultural Embeddedness of Engineering • The Internal Culture of Engineering • Engineering and Science
Developing and using models	• Design in Engineering • Specifications, Constraints, and Goals • Models of Design Processes • The Internal Culture of Engineering • Engineering and Science
Planning and carrying out investigations	• Design in Engineering • Specifications, Constraints, and Goals • Models of Design Processes • The Internal Culture of Engineering • Engineering and Science
Analyzing and interpreting data	• Specifications, Constraints, and Goals • Models of Design Processes • Engineering and Science
Using mathematics and computational thinking	• Design in Engineering • The Internal Culture of Engineering • Engineering and Science
Designing solutions	• Design in Engineering • Sources of Engineering Knowledge • The Scope of Engineering • Models of Design Processes
Engaging in argument from evidence	• Specifications, Constraints, and Goals • Sources of Engineering Knowledge • The Internal Culture of Engineering • Engineering and Science
Obtaining, evaluating, and communicating information	• Specifications, Constraints, and Goals • Sources of Engineering Knowledge • Knowledge Production in Engineering • Cultural Embeddedness of Engineering • The Internal Culture of Engineering

When students develop models for their potential solution to the problem, teachers can point out how students must understand the science behind the model. Students must understand how the variables, relationships, and mechanisms of the parts of the model behave in the natural world for the model to be a potential solution. The model for the solution also needs to fit into the specific requirements, constraints, and goals of the problem they are investigating. Teachers can communicate to students that design in engineering values parsimony and elegance, which must also address the specifications, constraints, and goals of the problem they are investigating.

Students then investigate the factors of the model they designed by creating and implementing the investigation and analyzing the data on how well the model addresses the problem. In doing so, teachers can point out that the way procedures are designed matter in the discipline of engineering. Testing a prototype only once or testing it outside of the constraints or goals of the problem is not valued in the discipline of engineering. Science is interrelated with engineering and with these practices as well because the laws of the natural world will govern how the model or prototype works to address the problem. Similarly, using mathematical and computational thinking is useful to engineers because this type of thinking is precise and logical, which makes the solution more elegant and more likely to be communicated clearly to the client.

Once a model has been tested, students then design solutions from what they learned from the tests and create an argument for why their solution addresses the problem with necessity and sufficiency. When they design solutions, teachers can point out that solutions in engineering can effectively solve problems but those problems have a scope and not all of the problems may be solved. The problem may be alleviated by the solution in part, but often the designed solution has implications for society. Teachers can also point out to students that the strength of an engineering design process helps develop stronger solutions for the identified problems. The arguments that students create about how the designed solution solves the problem must be grounded in the principles of science, otherwise the solution would not realistically work. Teachers can show students that engineers pay attention to the arguments of other engineers and how collectively the information gathered across many different investigations contributes to the body of knowledge in the different domains of engineering.

When students are consumers of engineering information by obtaining, evaluating, and communicating information regarding engineering problems, they can be reminded to be critical of the information using the culture and processes of engineering as a guide. Students can be aware of the scope of the problem, goals, and constraints in the information provided and critique the alignment. In the same way, students can follow the information for the strength of the engineering design processes used and evaluate the outcomes based on the norms of the discipline of engineering.

Engineering design processes (EDP) may be more familiar to educators than the nature of engineering. Teachers can use the engineering design cycle as a way to connect the discipline of engineering (epistemic knowledge) to engineering practices (procedural knowledge), although the design cycle leaves out some of the cultural aspects of the nature of engineering. Engineering practices comprise the implementation of the design cycle. Figure 2.1 displays an alignment of the engineering practices and the processes from engineering design. The engineering design cycle can be conceptualized as a method of investigation in engineering and the engineering practices are ways to carry out the method. Figure 2.1 lists engineering design processes that are associated with each engineering practice. Note that the practices overlap and can be used multiple times during the design process.

2.3 Why Are Science and Engineering Practices Difficult to Teach and Learn?

In the United States, science learning standards from the early 1980s to 2010 have focused on mainly content knowledge with only a few standards on inquiry and the nature of science. This may due to the fact that it is easier to assess outcomes than processes. In an oversimplified way, outcomes can be thought of as content knowledge and process can be thought of as procedural knowledge. Content knowledge is easier to assess because it is factual knowledge that has less divergence than in process knowledge. There are many different ways to conduct an investigation that yields valid and reliable information but often content knowledge in a science or engineering field is agreed upon and more convergent.

In the field of educational psychology, there are well-defined constructs called outcome and process goals. Outcome goals focus on a product, such as a display or presentation of content knowledge. Process goals focus on the way to go about getting to the product (Zimmerman & Kitsantas, 1997). Educational researchers have found from experimental design investigations that people who were learning the same task in three conditions (outcome goal only, process goal only, and process then outcome goal) performed best in the last condition. Given only an outcome goal, a person may struggle on their approach of how to get to the final product. Given process goals only, a person could have clear guidance about how to get to a product, but when they master the product, giving them more process goals is repetitive and demotivational. However, giving process goals only until a person masters the way to approach a task, then

Unpacking Science and Engineering Practices

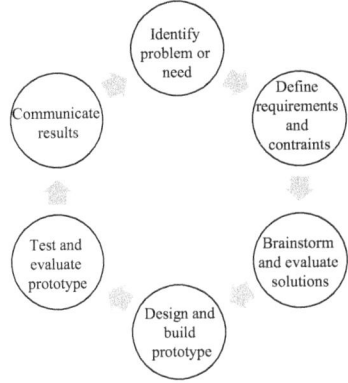

Engineering Practice	Connection to Engineering Design Processes
Defining problems	• Identify problem or need • Define requirements and constraints
Developing and using models	• Brainstorm and evaluate solutions
Planning and carrying out investigations	• Design and build prototype
Analyzing and interpreting data	• Test and evaluate prototype
Using mathematics and computational thinking	• Brainstorm and evaluate solutions • Design and build prototype • Test and evaluate prototype • Communicate results
Designing solutions	• Brainstorm and evaluate solutions • Design and build prototype • Test and evaluate prototype • Communicate results
Engaging in argument from evidence	• Brainstorm and evaluate solutions • Test and evaluate prototype • Communicate results
Obtaining, evaluating, and communicating information	• Brainstorm and evaluate solutions • Test and evaluate prototype • Communicate results

Figure 2.1   Connections between engineering design processes and engineering practices

switching to giving only the outcome goal tends to be an effective approach. In this circumstance, the learner knew how to get to the outcome and only needed to know what to produce. Science and engineering practices are, in effect, process goals if they are a way to complete an investigation to gain more knowledge.

Teaching process goals can be difficult to assess because they are inherently divergent. Teachers, even though they expressed value in teaching and assessing process goals, tended to measure only outcome goals even when they were given instruction on process goal assessment (Porter & Peters-Burton, 2021). Of course, this is not meant to blame teachers. It is a systemic issue in education. The content in this book focuses on process goals for science and engineering practices and should be considered the first few steps in a long and difficult journey of teaching students to think like scientists and engineers. The book is focused on teaching both process and outcome goals for beginners. Once students master the level of science and engineering practices that are appropriate for them, teachers should then shift their instruction to outcome goals and allow students to independently show their mastery of science and engineering practices. If students find that they cannot yet be independent in the science and engineering practices, then teachers can support them by helping them set, monitor, and reflect on the process goals that will help lead them to mastery of the practices.

2.4 What Are the Learning Tasks for Each Science and Engineering Practice?

In order to develop process goals to guide students to gain mastery of science and engineering practices, each practice must be decomposed to find the way to the outcome goal, which is being able to conduct the practice in the manner of a scientist or an engineer given different contexts. As teachers, we cannot support students by merely requiring them to "ask a scientific question." It is likely that students will not know the key characteristics of a scientific question. The characteristics of the task of asking a scientific question should be broken down and explicitly explained to students. Students should have a chance to practice their new skill under the guidance of the teacher and be given feedback about the level of their success in performing the practice. Additionally, it is helpful to decompose each science and engineering practice for the taxonomy of the types of tasks in the practice for the purposes of research.

Breaking down each practice can help with coding qualitative data or creating scales that look at all aspects of a practice. The remainder of this chapter explains the tool used to break down the science and engineering practices given in detail in Chapters 4 through 11.

2.4.1 Task Analysis Table

A task analysis table is a tool that helps to break apart teaching and learning activities (Table 2.3). The decomposition of teaching and learning activities gives a teacher the ability to look closely at where the science and engineering practices are located in the lesson, if they are appropriately placed, and if there are opportunities for more practices in the lesson. The task analysis table consists of a three-column matrix as shown in Table 2.3 (a partial example of a lesson plan). The teacher examines their lesson plan and then captures the essence of what they are doing to facilitate the lesson and what students are doing to learn. Sometimes the "Teacher does" and the "Student does" entries need to be adjusted to the right level of detail by the teacher to be useful.

Task analysis tables have been used in the past to align assessments to learning goals (Feldon et al., 2010). Task analysis tables have also been used to help teachers integrate computational thinking practices into data practices (Peters-Burton et al., 2022). In this study, the high school science

Table 2.3 *Task analysis table with an example activity*

Teacher does	Student does	Science and engineering practice
Ask the class what they think will increase the strength of an electromagnet		Asks questions
	Identifies parts of an electromagnet to determine the variables in the phenomenon	Developing and using models
Organizes groups of three students to write a procedure to test the variables		
	Students collaborate to write the procedure to test a selected variable from their model	Planning and carrying out investigations

teachers who were learning to use the task analysis table while integrating computational thinking into their science investigations reported that "being able to identify that the lesson has those components is really empowering," and "I found it [the task analysis tool] really useful to think about how I guide kids step by step through a process. I don't think I had really thought about some of the questions that I asked them for every single step. I think so much of what we do becomes automated because we're just used to doing it, and we don't stop and think."

The use of the task analysis table had an unintended benefit for the teachers when we used it in the study. After the teachers reviewed their lesson plans and filled out the "Teacher does" and the "Student does" tables, some of the teachers noticed that they were performing the science practices and the students were just reacting. Before the teachers attempted to integrate computational thinking, they took the opportunity to make the lesson more student-centered.

When using a task analysis tool to identify science and engineering practices, teachers can easily determine if they are doing the science and engineering practice for students or if the students are doing it. We typically want to have student-centered classes, but there are times when it is appropriate for the teacher to model the practice before the student attempts it. When students are first attempting a science and engineering practice, they often need the teacher to model it so that they can see the outcome goal. For example, in a lesson on electromagnets, the teacher asks the question about what variables change the strength of an electromagnet. The students attempt the other practices in the example (modeling and designing a procedure), but the question is given to them by the teacher. However, teachers can further support their students by also modeling the process goals that will help them accomplish the outcome goal. Chapters 4 through 11 of this book are organized to help teachers identify key characteristics in successfully performing the practice.

Appendix F in the Next Generation Science Standards articulates the key characteristics for each science and engineering practice for grade bands K-2, 3–5, 6–8, and 9–12. These key characteristics can be the foundation of process goals that teachers can model for students, then help students to use those process goals themselves, eventually allowing students to be able to understand and perform the outcome goals without assistance. For example in the K-2 grade band expectations for the practice of asking questions, the outcome goal is for students to be able to create simple descriptive questions that can be tested. In order to progress to that goal, students can ask questions about the natural phenomenon so they can understand it more clearly. Another process goal would be to understand what changed in the phenomenon. Finally, students can set a process goal to identify what can be tested from the phenomenon. Once they have mastered these three process goals, they can likely form a question about the phenomenon that can be tested and accomplish the outcome goals. Each of the practices in Chapters 4 through 11 break down the outcome goal into process goals so that teachers can model the practice for their students. Later teachers can use the process goals to guide their students as they take part in the science and engineering practice. Finally, when the student masters the practice, they are able to perform the outcome goal with no outside assistance and can transfer their learning to other contexts.

CHAPTER 3

Self-Regulated Learning

3.1 What Makes a Self-Regulated Learner?

As science educators, we know that student-centered learning is the gold-standard of how to teach students about science, because student-centered learning has the ability to engage and empower students to learn science on their own. One of the most powerful ways to help students become lifelong learners in science is to teach them disciplinary skills, such as science and engineering practices. When students know how to use practices in a way in which they can obtain valid and reliable knowledge, they can learn about the world around them outside of science class. Students who know how to use reliable resources to find out how one variable influences another have the tools to be science-minded when they solve problems. The purpose of this chapter is to demonstrate to teachers tangible ways in which they can integrate science and engineering practices supported by self-regulated learning theory into their science instruction so that students become more aware of how they learn these skills. Students can then possess scientific and engineering ways of knowing to add to their other ways of knowing.

During my time as a secondary science, engineering, and mathematics teacher, the notion of teaching students how to think like a scientist, engineer, and mathematician was murky for me. I wanted my students to be good observers, but what did that really mean? How could I teach my students to pay attention to the relevant observations and not be distracted by the irrelevant ones? How could I get my students to respect evidence and not always choose the red car in a physics investigation regardless of the other conditions because they think red cars are faster?

I have been fortunate later in my career to clear the murky ideas behind teaching practices by being involved in educational investigations with middle school students and secondary teachers. As researchers, we aren't anywhere near our final destination, but we have made some progress in being able to explicitly develop scientific and engineering thinking. I'd like to report a comment that one of the students participating in an intervention said to me in order to demonstrate the purpose of this book. "I knew that I was supposed to be thinking SOMEHOW in science, but I didn't know how. But these types of lessons showed me exactly how I should be acting and thinking like a scientist." Honestly, a thirteen-year-old student who didn't know me well came up to me and said this! This quote motivates me to work with teachers to continue developing students' scientific thinking and I am fortunate to be here to share some of this research-based practice. I feel that the ideas and examples in this book will help teachers design some of their own ways to help students see the world through science and engineering.

3.2 Mechanisms for Learning Effectively

Before we can get to learning science, we need a mechanism for learning. I promised myself that as a teacher educator I wouldn't just tell teachers to "just do it" and continue presenting the "black box" for teachers in my professional development settings. When I was a high school track coach, the coaches from the other sports sometimes made fun of me. They said, "How hard is it to be a track coach? You just tell your runners to run faster!!" I promise that although learning processes in this book may seem complex, I will definitely NOT tell you to teach faster, teach harder. The learning processes and strategies presented in this book, which are backed up by empirical educational research, will help you to teach science deeper because you will be reaching out to students and teaching them how to learn and how to think efficiently. In doing so, you will show your students that you care about how they build their knowledge so that they can prosper and be lifelong learners.

In the same vein as a track coach, you can't go to your classrooms and tell the students "Learn faster, learn harder." Even older students need support in learning how to learn. This is where self-regulated learning comes in. Self-regulated learning helps students learn how to be aware of their learning, which gives them ownership. This in turn motivates students to continue to monitor their learning and helps them to reflect on how well they did and learn from their mistakes. Knowledge about self-regulated learning processes helps students to make improvements in their learning processes. It helps students to understand it is okay to make mistakes while learning, but also to know where they struggled and to act on possibilities of fixing unhelpful processes the next time they learn.

3.3 What Is the Theory Behind Self-Regulated Learning?

Self-regulated learning is a theory that explains how learners can be metacognitively aware of their learning, notice patterns of strategies that work and that do not work for them, and adjust accordingly. Self-regulated learning has been studied for over forty years and has worked to improve teacher and student learning across a variety of ages of learners and topics of learning. There are three phases in self-regulated learning; it is helpful to think about them in terms of (1) *before learning,* (2) *during learning,* and (3) *after learning.* For SRL processes to be productive, the learning task should be well defined so that students can set goals, monitor their goals, and reflect on their outcomes. If the learning task is nebulous, it is difficult to know if you have met your goals or not. Figure 3.1 provides the overview of the processes of learning in self-regulated learning theory that will be explained in the next section.

3.3.1 Forethought Phase – Before Learning

A learner comes to a problem with what they already know, their experiences, in this case science and engineering practices. Often this includes what they have already learned about science or engineering and experiences they have outside of class about the natural or designed world.

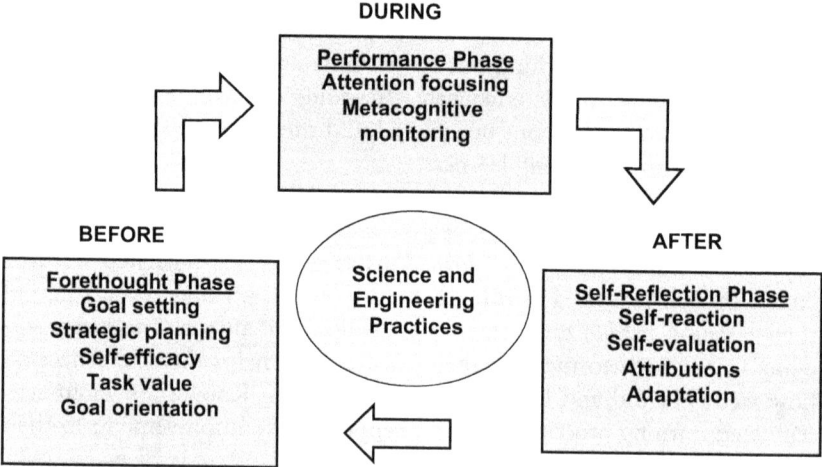

Figure 3.1 Phases and sub-processes in self-regulated learning theory
Source: Adapted from Zimmerman (2000).

During the forethought phase, a self-regulated learner would *set goals*, preferably both process goals and outcome goals, and set up a *strategic plan* regarding the specifics of how to get started and remain on track to meet the goals. In addition to being motivated to set goals and a strategic plan, a learner comes to a learning task with beliefs and orientations. They have a level of *self-efficacy* for the learning task based on prior experiences with mastery, watching others do the learning task, encouragement from others, and the emotions they have about the task such as excitement or dread (Bandura, 2002). Having a high self-efficacy is important because learners who have a high self-efficacy tend to take more academic risks and are more resilient when they fail (Perry, Phillips, & Hutchinson, 2006; Zimmerman & Kitsantas, 2014). Learners also approach a learning task with a level of *task value*, which is their perception of how important learning the task is to them (Deci, 1975). A productive self-regulated learner finds value in the task in order to self-motivate to accomplish the task successfully. A colleague and I found in a study of elementary teachers, that even if teachers had a low self-efficacy for learning how to teach earth science through inquiry methods, but if they had a high task value, they persisted and were eventually successful in designing and teaching earth science through inquiry methods (Peters-Burton & Botov, 2017). We also found that teachers who were able to master the learning possessed higher self-efficacy at the end of the professional development program. Another belief that learners have when they approach a learning task is *goal orientation*, which is a continuum of beliefs that range from learning for mastery to trying to get through the learning task looking successful (Ames, 1992). A productive self-regulated learner has an orientation toward mastery and isn't afraid of failing publicly. Rather, they are ready to learn from failures and to be persistent in the learning task. Armed with these prior beliefs and orientations and with goals in mind, a learner proceeds to the performance phase.

3.3.2 *Performance Phase – Actively Learning*

In the performance phase, a learner enacts content, procedural, and epistemic knowledge that they possess to accomplish the learning task. As a teacher, you may have already had strategies to support students in enacting relevant knowledge, for example, with a Know-Want to know- and have Learned (K-W-L) chart. The K-W-L graphic organizer helps students think about and write down the knowledge and skills and beliefs they have related to the learning task. The chart helps them list relevant ideas learned in school and in informal settings. In order to be efficient, self-regulated learners *focus their attention* on relevant ideas that will help them meet their goals. We know from the literature on expert and novice learners that novice learners often focus their attention on extraneous things because they do not have epistemic knowledge or see the big picture of why they are doing what they are doing. Often, there are students who have not yet experienced science and engineering in a deep way prior to entering classrooms. Therefore, these students may not know the key factors on which to focus, particularly in science and engineering practices because they can be so divergent in nature. One of the goals of this book is to make explicit the key factors in learning science and engineering practices, which will assist teachers and teacher educators in focusing student attention on the most important points of doing science and engineering in classrooms, even those who have not yet experienced science or engineering in other settings. In order to focus attention in a productive way, learners can refer back to their goals. Similarly, learners can *self-monitor* their progress in the performance phase by checking their work with their goals to see if they are reaching the process goals in order to lead to the outcome goal. Teachers can help students by modeling *self-monitoring* and then reminding students to monitor themselves at strategic times in the learning task.

After learners do the learning task in science class, then they have outcomes of learning. Often in the classroom, it is difficult to take the time to reflect. It may feel like in order to accomplish the curriculum in full, the class must move forward. However, in self-regulated learning, the cycle isn't complete until a learner reflects on their learning to assess their performance. Giving students time to reflect on what worked and what did not work for their learning is worth the investment in time. Eventually, students will become more effective and efficient learners, which will ultimately save time.

3.3.3 *Self-Reflection Phase – After the Learning*

After learning, a learner checks the outcomes with a standard, similar to the way that teachers check that lesson plans address the standards of learning. In self-regulated learning, a learner checks their performance against standards to see if they did well or not and address what to change about their learning processes in the future. When a learner receives feedback, whether that is physical feedback from an investigation, feedback from another person like a teacher, or self-feedback, they first *react* to it and accept the feedback in part or in whole or reject the feedback. If they accept the feedback, then they *evaluate* what went well and what did not go well in meeting the goals for the learning task. They reflect on the process of learning and *attribute* their successes and failures to different things. Some attributions are more productive than others. For example, if a learner attributes a failure to write a procedure for an investigation that isolates a variable to the notion that they were never "good" in science and their parents weren't either, then that is not a productive attribution because it is unchangeable. However, if a learner attributes a failure to write a valid procedure for an investigation on their being distracted by an out-of-school event, then that is a more productive attribution because it is changeable. The learner noticed the reason for the flaw and took responsibility for it. Note that the learner did not think they personally were a failure because they had a failed learning event. Similar to engineering, when you test an idea, there is bound to be some level of failure, but the important part of an engineering design process is that there is information gained from the failure to apply during the next attempt. When a self-regulated learner takes account of things that went well and things that did not go well, then they *adapt* what they did not do well by seeking out other strategies that might work the next time they learn in a similar way. Seeking help might take the form of looking-up for information about learning processes that might be pertinent, talking to peers who were successful in this type of learning, or asking a teacher what they might do to improve.

Self-regulated learners then apply this new awareness of their learning processes into the next forethought phase when they attempt another learning task. Self-regulated learning is cyclic. Each attempt at a new learning task can build up awareness and skill in learning.

3.4 Examples of Self-Regulated Learning

The act of serving a volleyball is introduced in this section to initially illustrate the cycle of self-regulation in action. Imagine you are on a volleyball court about to serve the ball. When you are standing there holding the ball at the back of the court, you think about what you need to do in the near future to get the ball over the net in a way that your opponent will have difficulty returning. You organize what you know for your future performance. When you become more experienced, these organizational thoughts become more automated and less intentional. The performance phase occurs when you hit the ball away from you.

The self-reflection phase happens quickly after that. If the ball goes over the net and your opponent has difficulty returning it, then you have successfully met your goal. You remember what you did that made it go well and if you are a good learner, you "catalog" it for future use so you can do it again. This gets stored into your forethought and you have a more sophisticated forethought for the next time you serve the ball.

You can think of the same type of scenario with asking scientific questions. Since questions must be testable and ultimately have descriptive power, in the forethought process an effective learner would think about their model of the phenomena and identify areas in which to ask questions. This helps them organize what they already know and to set goals for extending their learning about the phenomena. The performance phase occurs when the learner structures and refines the question. Then when the learner gets feedback on the question, either from a teacher, a peer, or from trying to answer the question themselves, they consider their performance against the feedback and react. If they are effective learners, they see how their question was productive in extending knowledge and what made the question appropriate for asking about the phenomena. The cycle of self-regulated learning can be most helpful if learners are aware of their processes and act to improve their learning. In the next section, I will explain a case study of a student who is not yet proficient in self-regulated learning and a case study of one who can effectively self-regulate.

3.4.1 A Naïve Learner

A naïve learner, we will call her Natalie, begins by setting non-specific, distant goals such as "I will get an A in science." Natalie has set a goal, but it is not a productive or informed goal and will be difficult to monitor and assess during her performance in science class. Natalie has decided that she wants to be a high performer in science, no matter what, and is less interested in the subject matter than she is in getting on the honors list. Outwardly this may sound like a goal that will drive Natalie toward success, but that success is not focused on learning and can possibly be attained by memorizing material for assessments. She wants to do well this year in science, but she hasn't done well in science in the past and she cannot coherently express the plans that she has to change her learning strategies. She knows that science operates differently than other subjects, but cannot describe the "rules" of scientific inquiry, which makes her less interested in pursuing her studies of science outside of class. When Natalie goes to science class and performs her tasks in class, she does not have a focus beyond getting the assignments completed. Often, she will enter into an investigation by saying to herself, "I don't really know how to be a scientist, so I won't be surprised if I fail this lab" which sets up her attribution for failure to be an unchangeable, innate entity.

She only knows if she is doing well in science when her teacher gives her a grade or some verbal feedback, which Natalie avoids as much as possible because she just wants to complete the assignments. When she fails in a science assignment, she resigns herself that she just isn't scientifically minded, so she doesn't try anything differently to learn how scientists think. As a result, she continues to fail at learning science beyond a superficial level.

3.4.2 An Effective Learner

A skillful learner in science, Sabrina, begins by setting an outcome goal for the year that is focused on learning rather than on recognition ("I will learn the role of evidence in science") and sets process goals as benchmarks to reach this distant goal ("I will first learn why many trials are needed in investigations, then I will learn what is meant by valid results"), and she understands the hierarchy of her set of goals. This process of setting goals has worked for her in other classes, so she believes that this strategy will work for her in science class this year. She feels ownership of this type of learning because the teacher has explicitly described how scientific epistemologies relate to the content knowledge and to the labs they do in class, so she is intrinsically interested in learning how to think scientifically. She feels comfortable being challenged in her scientific thinking because the teacher supports her synthesis of domains of knowledge by pointing out how different fields of science discover and validate knowledge.

During class, Sabrina often asks why particular processes in science are used, and the teacher elaborates on the scientific enterprise in response. Sabrina is not afraid to show publicly if she does not know something in class. She is aware of what she needs to do to solve problems in science because instead of following a prescribed set of steps, Sabrina has developed a way of knowing in science that allows her to compare how professional scientists act and think to her own thinking. She actively seeks out help when her ways of knowing in science don't correspond to the nature of science. She doesn't worry too much about "messing up" during science labs, because she knows that the teacher provides students opportunities to reflect on the logic of their own thinking.

When Sabrina fails, she knows that she needs to try some other learning strategy, and she draws from the explicit instruction the teacher gives her about scientific inquiry in class. The compilation of successes and failures along with their corresponding learning strategies help Sabrina to make a library of academic strategies that work for her. She draws from this library of academic strategies to create new goals, with the help of the teacher's instruction.

3.5 What Can Teachers and Teacher Educators do to Support Student Self-Regulated Learning?

This section will describe some brief and general examples of how teachers and teacher educators can support student self-regulated learning. I'm hoping that these general examples will inspire you to keep progressing through the book. Chapters 4–11 give more specific examples for each science and engineering practice. Phase-wise examples from the SRL cycle are given below.

3.5.1 Forethought Phase Examples

In the forethought phase, self-regulated students take into account what they already know about the learning task and their beliefs. They independently set goals for themselves to accomplish the learning task. Teachers can support students who are not yet self-regulated by modeling strategies for students in two ways: (a) how they approach and manage their own beliefs when getting ready to learn something new, and (b) how they set goals and strategic plans to meet their goals.

Teachers can support students by showing them instances of approaching a learning task where they had high self-efficacy and instances where they had low self-efficacy. In the case of high self-efficacy, the teacher can explain how they obtained these productive beliefs such as having tried to learn something similar, and they succeeded and cataloged what they did well. In the case of low self-efficacy, the teacher can explain to the students how, even though their level of confidence for this particular learning task was low, they talked themselves into trying something new, and reasoned that if they failed entirely, they would still learn something new from the experience. They may have also sought out someone they trust and can relate to in order to see how they approached this type of learning task in order to boost their self-efficacy.

Teachers can also help students to be more aware of how task value influences their motivation to follow through on the learning task. First, teachers can be explicit and ask students to rate how much they value a learning task and why they gave that rating. Students can have high task value for a learning task and explain that they see how these skills they are about to learn are applicable in many ways in their lives. Teachers can support students who have a low task value for the learning by finding out why they have a low task value and showing how the learning is valuable from the student's perspective. For example, perhaps the student has low task value for using mathematics and computational thinking. The teacher can find out what the student may want to do in the future for a career and give examples of how mathematical and computational thinking is crucial for that career. The teacher could also boost student task value for learning about mathematical and computational thinking by illustrating things that help adults to be more prosperous, such as using a spreadsheet to maintain a budget and predict inputs and outputs of resources. Skill in breaking down ideas to put in the spreadsheet and setting up formulas to automate reoccurring expenses could capture the interest of the student who may have not made the connection between the classroom work and future opportunities.

Setting goals and creating a strategic plan to meet those goals is crucial throughout the cycle of self-regulated learning. Setting goals hinges on the success of attention focusing and self-monitoring in the performance phase and in self-evaluating and attributions in the self-reflection phase. Teachers can support student goal setting by helping students understand the difference between outcome and process goals, and that when students are learning something new they should set process goals as small steps that lead up to the outcome goal. Teachers can model goal setting for students to demonstrate their learning processes, keeping in mind that a process that might work for one person might not work for another person. Teachers can also teach students advantageous goal-setting strategies (Zimmerman, 2008).

When goals are more specific, they are more attainable. For example, a specific goal regarding learning about science would be "I will be sure that each of my assertions in my conclusions are directly related to evidence I observed," whereas a general goal would be "I want to do science better." The specific goal is more distinct and measurable, and the learner is able to determine when they have achieved the goal. The general goal is too nebulous to be measured, and the learner will be unsure when the task has been successfully completed. Teachers can help students create SMART goals (specific, measurable, attainable, relevant, and timely) as a way to design their goals to be specific. Students need practice in creating productive goals, so teachers can spend the first month of school practicing goal writing daily, with daily feedback to each student. Create a goal setting form where students have to report the specific information they are working on that day and report on the specific strategies they are trying that day.

Goal setting should also include proximal and process goals. Proximity of goals refers to the length of time in which the goal is intended to be reached. An example of a proximal, process goal in the short term would be "I will understand the relationship between velocity and acceleration in physics by the end of the week," as opposed to a long-term, distal, outcome goal, "I will learn all of the major theories in physics and their role over time in developing the field of physics by the end of the year." Proximity of goals is related to the hierarchical organization of short-term process and long-term outcome goals. A proximal, process goal provides ways to remain on task and to produce a successful outcome than a distal, outcome goal does. If the process and outcome goals are properly organized into a hierarchy that the proximal process goals must be reached first, can help the student attain the distal outcome goal. If only distal outcome goals are set, the learner will have more difficulty being motivated to reach the goal than if the learner set both proximal process and distal outcome goals.

Congruence of goals with classroom learning is important. That is, if the goals match what is valued as learning, then the goal is easier to reach. Learners who have congruence of goals regarding science would aspire to understand how knowledge is acquired and validated in the scientific enterprise, and would not be deterred by conflicting goals set from external sources, such as a classroom oriented only toward memorization of scientific facts. If a goal is not aligned with what is valued in school, then the goal and the environment in which the learner is existing do not match and attaining the goal is much more difficult.

Finally, goals should be set in a zone of development, not too easy to reach and not too hard to reach. Goals that are set higher than can be achieved are not productive. Learners must be able to set goals that are challenging, yet within their reach. For example, take a student who has below grade level reading skills. Science textbooks are notoriously high in reading level, often three grade levels above the grade for which they are written. A goal for this student at this time that may be too challenging would be to read and understand everything in chapter 2 in the textbook they are using. However, the teacher can guide a student to set a more attainable goal such as, to read all of the captions and look at all of the illustrations and graphs in chapter 2 and predict what the chapter is about.

Many times, students do not have a great deal of experience with science and engineering outside of school, so it may be difficult for a novice learner to set their own goals. Teacher support is necessary to help learners set goals that are reasonable in a field where they do not have much experience. For example, a learner who wants to understand about the inherent guidelines from which science as a discipline operates and is in a learning environment that explicitly points toward nature of science examples within the context of the classroom learning has congruence of goals, the student's goals and the environment match. An example of how science instruction parallels this principle is the breakdown of the whole of science into topics such as motion, energy, reactions, cell theory, and the like. Learning the whole of science would be overwhelming to a learner if it were not described in more reduced terms.

Teacher educators can help teachers become aware of all of these opportunities to teach students how to learn effectively. They can form professional development communities for teachers to share their experiences in supporting student self-regulated learning.

3.5.2 Performance Phase Examples

Teachers can support students by focusing their attention to the key features of the science and engineering practice and by modeling how to monitor progress toward the process and outcome goals. Before teacher support can happen, teachers must break down what the key features are of each practice so that they can show students the specific skills that make up what is valued in the practice. Later chapters in the book break down key features for each practice, but since this is generalized for a larger audience, teachers will need to translate this information into their own context and teacher practice.

As an example of breaking down key features of a practice, let's examine how a student might approach creating an appropriate graph to display data in order to communicate findings. Before making a graph students will need to do the following:

- Identify variables and units
- Decide what type of graph (bar graph, continuous line graph, histogram) will best display the type of data gathered
- Decide how many trials, central tendency, and variance will be displayed
- Title and axis

After making a graph, students can ask themselves:

- Does it make sense?
- Should I go back to do a redo?
- Is there enough data? Not enough?
- Are the title and axis labels accurate?

Teachers will need to consider the appropriate developmental level for these skills and on the context of the learning to decide what to model for their students.

Modeling self-monitoring for students can help them become more aware that being mindful while learning will help them become more self-regulated in learning. Teachers can also review the progression of an assigned investigation to find optimal places to remind them to monitor their progress. Students can do so by checking to see what process goals they have accomplished to monitor if they are progressing toward their outcome goal. Teachers can formalize this for students who are new to the process by creating a form on which students write their goals and have space to journal their progress toward process and outcome goals. Teachers can fade their support once they feel students do not need the modeling or the reminders to self-monitor. They can trust that students will fill out the form by themselves and move toward a more automatic self-monitoring of goals that may not require a written form in the future.

3.5.3 Self-Reflection Phase Examples

In the same way that teachers can support performance phase self-regulated learning processes by modeling and reminders, teachers can support student self-reflection phase processes. First, teachers can be explicit about obtaining feedback and relay to students the importance of feedback as information. If students think feedback is only about evaluation, they may learn to dread feedback because they see it as a personal judgment. Rather, teachers can create learning environments where students seek out feedback in order to improve their learning processes.

When students receive feedback, teachers can remind students that they should take feedback as information on their learning processes, not on their personal characteristics. These reminders may assist with students' self-reaction to feedback and allow them to not have an immediate, defensive reaction but instead take in the feedback for improvement. Teachers can help students self-evaluate feedback by showing students how to look at details that make up valuable feedback. For example, feedback that has only good/bad judgment is not helpful for improvement. Feedback that has specific context details for the learning task can be useful for students.

Teachers can also help students with attribution of successes and failures by reminding students that it is their *processes* that lead to success and failures, not innate characteristics of a person. Everyone can learn to be a better learner. Teachers can also point out types of attributions that can be helpful. External, uncontrollable sources of attribution such as "I failed because the teacher is too hard" is not productive. Internal, controllable sources of attribution such as "I failed because I just didn't spend the time preparing" is more productive to a learner because they learn that preparation is needed the next time. In order for attribution to be helpful, students must be honest with themselves. This tends to be out of the hands of teachers, but frequent modeling by the teacher about the ways they attribute their successes and failures to their learning processes may encourage students to be more self-reflective.

Once productive attributions are identified, self-regulated learners are also able to adapt their unproductive learning processes. This may be in the form of trying something new the next time they learn, such as trying to find more value in the task or self-monitoring more frequently. It may also be in the form of help-seeking. When students are at a loss to replace an unproductive learning process with a new learning process, they can ask a peer or a teacher what they might try. They might also do some research into learning processes and attempt a new approach from external information. Teachers can help students be more adaptive learners by demonstrating a variety of learning strategies so that students can have many examples from which to try.

Teachers' actions are extremely important for supporting student learning in science and engineering. Students may not have had detailed prior experiences with science and engineering. As a result, students will need the teacher to model the practices for them so that students can observe what needs to be done to master this type of procedural knowledge. Students may not have had any experience with self-regulated learning and would benefit from teacher modeling of these processes as well. Teachers are the experts in both content and learning processes. In order for students to move to mastery learning, teachers can help students by demonstrating the process goals that will lead to mastery outcomes in science and engineering practices.

3.6 Coaching for Self-Regulated Learning in Science and Engineering Practices

Since students cannot be expected to innately self-regulate their learning, teachers play a vital role in supporting student self-regulated learning of science and engineering practices. Zimmerman proposed a coaching structure for supporting self-regulated learning that consists of four phases: observation and modeling, emulation, self-control, and self-regulation. There has been prior work with this coaching structure that has demonstrated that it can be effective generally in learning (Zimmerman, 2000) and specifically in science and engineering practices (Peters, 2012; Peters-Burton & Burton, 2020). Educational researchers can explore the coaching structure as a whole or in parts to further explain the effect of coaching on students, how teachers experience this coaching method, and the interactions between teachers and students that lead to outcomes. In the *modeling* phase, the teacher pinpoints the key characteristics of a particular science and engineering practice and overtly performs them for students while talking aloud about their thought processes. Students should note the approaches in the practice. When teachers feel that students can identify the key characteristics of a practice, teachers can give students a chance to *emulate* the approaches to a science and engineering practice by reminding students of the key features they modeled and providing feedback to the student on their attempt based on these key features. When teachers feel that students have sufficiently emulated the science and engineering practice with teacher support, then teachers can fade some of the key features of support for the science and engineering practice and provide only minor support to students. When students are successful in accomplishing the science and engineering practice with minor teacher support (modeling and reminders of key characteristics of the practice) then teachers can fade all support. If students can perform the science and engineering practice to the level of the teacher's expectation by themselves, then they can be considered a self-regulated learner because they have learned to do the practice independently.

Although we know some things about how teachers use self-regulated learning to understand content knowledge and pedagogical content knowledge (Cleary et al., 2022; Porter & Peters-Burton, 2021) and how students use self-regulated learning to learn science and engineering practices (Peters-Burton, 2015), there is much more to know. We do know from multiple studies that learning how to self-regulate one's own learning takes a great deal of time (Cleary et al., 2022; Porter & Peters-Burton, 2021;

Tran, Capps, & Hodges, 2022). We also know that self-regulated learning can be used to help students learn content knowledge (Peters, 2012; Peters-Burton, 2015), procedural knowledge (Peters & Kitsantas, 2010), and epistemic knowledge (Khishfe & Abd-El-Khalick, 2002; Peters-Burton, 2015) in science.

In addition to investigating teacher and student interactions and types of support in self-regulating science and engineering practices, researchers can investigate optimal approaches and differences in the types of learning tasks while students are learning to be self-regulated learners. Are there different self-regulated learning strategies for each type of practice? Is there a taxonomy of key features to accomplish science and engineering practices in individual practices or groups of practices? One might speculate that there are similarities in approaches to learning how to communicate findings and creating an argument using evidence because of the emphasis on language and the role of evidence. Understanding what the key features of science and engineering practices are from empirical studies and if there are more efficient ways of self-regulating the learning of particular science and engineering practices will help to move the field forward and improve educational experiences for all students.

PART II

Engaging in Disciplinary Tasks in Science and Engineering

CHAPTER 4

Asking Questions and Defining Problems

Chapters 4 through 11 in this part of the book have a parallel design. The practices of asking questions and defining problems are dissected into distinct and clear learning tasks and process goals are described for the practice. These tasks are then examined within the context of a self-regulated learning cycle and coaching strategies for this practice are explained in the MPI-S coaching model. Points for instruction and assessment are emphasized in the context of a design challenge for snack packaging targeted at students in grades 3 through 5. The snack packaging design challenge is then examined in two case studies – one positive and one negative – to demonstrate how the learning tasks can be used by students and how the teacher can support students for this practice.

4.1 Learning Tasks in the Practice

In the Next Generation Science Standards (NGSS), the overall definition for asking questions in science is to ask and refine questions that address the natural world and can be empirically tested. The overall definition of designing solutions in engineering is to clarify problems about the designed world by identifying criteria for solutions and constraints. Both scientists and engineers ask questions in their related field. These overall definitions have been further broken down into grade bands by NGSS of K-2 (aged 4 to 7), 3–5 (aged 8 to 11), 6–8 (aged 12 to 14), and 9–12 (aged 15 to 18). The following sections of this chapter explain each new objective for the science and engineering practice of asking questions and designing solutions for the grade bands. The sections begin with the learning tasks from K-2 and build on the new practices that are to be learned in the successive grade band (NGSS@NSTA, 2014). The learning tasks that are repeated from the prior grade band are not listed in the new grade band. From this foundation, the learning tasks are decomposed into what students should be able to do to master the practice.

Breaking the practice down is not only important for helping students who may not have had exposure to doing science and engineering, but it is also important to clearly define the learning task in a cycle of self-regulated learning. In order for self-regulated learning to be productive, one must begin with a well-defined learning task. As explained in Chapter 3, goal setting is key to being successful with self-regulated learning, and goals need to be discrete and composed of smaller, more proximal process goals that lead up to the outcome goal. In the overall definition of asking questions and defining solutions, some of the key characteristics of asking questions in science, such as focus on the natural world and the ability for questions to be testable, are apparent. Similarly for designing solutions for engineering, key characteristics include a focus on problems in the designed world, and an assessment of criteria and constraints of the problem. The individual skills are process goals, which together comprise the practice of asking questions and defining problems, which can be considered the outcomes goal. Students new to these practices may need process goals that are smaller steps to reaching the outcome goals.

The sections in this chapter also provide examples of how a teacher can support the learning task in an explicit way and in a reflective way. Educational research on epistemic knowledge in science has shown that students are less likely to learn an aspect of the nature of science implicitly, but more likely to learn that aspect when taught explicitly by the teacher and given intentional opportunities to reflect on how the aspect was demonstrated (Khishfe & Abd-El-Khalick, 2002; Peters, 2012; Peters & Kitsantas, 2010; Peters-Burton, 2015). The ideas from this body of literature are applied in a classroom setting to offer practical ways to enhance student learning for science and engineering practices.

4.1.1 *Process Goals for Asking Questions and Defining Solutions*

In grades K-2, the NGSS articulate the expectation that students will build on prior experiences and progress to simple descriptive questions that can be tested for the practice of asking questions and defining problems. Student learning tasks for that standard are broken into the process goals below:

- Recall and organize prior experiences related to the phenomenon
- Add observations related to the phenomenon to prior experiences for sufficient background knowledge

- Ask questions about the phenomenon that can be reasonably tested
- Define a problem that can be solved through the development of a new or improved tool

4.1.1.1 Explicitly and Reflectively Supporting Process Goals for Asking Questions and Defining Solutions

Educational research has demonstrated that students can more effectively learn when teachers explicitly and reflectively support learning tasks (Khishfe & Abd-El-Khalick, 2002; Peters-Burton, 2017). Note that in this research, explicit teaching is not lecturing to students, but is focused on teachers taking opportunities to point out to students when they are thinking and behaving like scientists and engineers. Teachers can explicitly teach by modeling the behaviors that are done by scientists and engineers when performing the practice of asking questions and defining solutions. When reflectively teaching the practice, teachers give students a chance to independently show what they know about performing the practice. After assessing the student's reflective response, teachers can determine if they need to reteach the practice or give students another chance to emulate the practice.

Teachers can explicitly support students to *recall and organize prior experiences* related to the phenomenon by engaging students with the general phenomenon for which they will be asking questions or defining solutions, providing a graphic organizer to students to record their prior experiences, and explaining to students that gathering background knowledge about the phenomenon is an important beginning to asking questions and defining problems. Teachers can reflectively support the learning task of recalling and organizing prior experiences related to the phenomenon by asking students to record prior experiences related to the phenomenon in the beginning of a unit of study. Then teachers can ask students to reorganize their recorded experiences on the next day, adding to and rearranging their ideas based on the new work. Students can reflect on their learning by explaining the purpose of recording their prior experiences. If students are able to reflect on the idea that background information is needed to ask questions and define problems, they have a sense of the rationale for performing this practice and have likely mastered this level of the practice.

Teachers can explicitly support the student learning task of *adding observations related to the phenomenon to prior experiences*. Provide students access to the focus phenomenon through hands-on equipment, demonstration, or video, and point out to students what might be of interest in terms of variables or relationships in the phenomenon. Teachers can reflectively support students in adding observations related to the phenomenon to prior experiences by asking students how their observations help them form questions and define problems. Students who understand why they do the practice may respond that knowing more about ideas being explored helps to create more specific questions and focus the problem.

Teachers can explicitly support the student process goal of *asking questions about the phenomenon that can be reasonably tested* by asking students what they might measure to answer their question or solve their problem, and pointing out to students that science and engineering requires information that everyone agrees upon (measurement). Teachers can reflectively support students in asking questions about the phenomenon by having students review other questions or problem definitions and explain if they can be tested or not and why.

Teachers can explicitly support student learning about *defining a problem that can be solved through the development of a new or improved tool* by asking students to list what they have as resources and what are the constraints of the problem. Point out to students that in order to define a problem, engineers need to know what resources they have to work with and what is out of bounds for the problem. Teachers can reflectively support student understanding of defining a problem that can be solved through the development of a new or improved tool by asking them what they need to know in order to define the problem. Students should be able to explain both the resources they have to work with and what is out of bounds for the problem in order to define the problem.

4.1.1.2 Explicitly and Reflectively Supporting Process Goals for Asking Questions and Defining Solutions for Grades 3–5

According to the NGSS, asking questions and defining problems in grades 3–5 builds on K–2 experiences and progresses to specifying qualitative relationships. A breakdown of student learning tasks that can serve as process goals are:

- Ask questions about what would happen if one variable would change
- Distinguish between qualitative and quantitative relationships
- Distinguish between scientific and non-scientific questions

Teachers can explicitly support students *asking questions about what would happen if one variable would change* by creating a model or hands-on apparatus of the phenomenon as a demonstration. The teacher can then ask students to identify variables and then point out when one changes how to observe the others. Teachers can ask students to reflect on their performance of asking questions about what would happen if one variable would change by first creating a model or hands-on apparatus of the phenomenon as a demonstration. Then have students identify variables for the phenomenon individually or in small groups, and have students ask questions about changing one of the variables.

Teachers can explicitly support student performance of *distinguishing between qualitative and quantitative* relationships by pointing out to students the pros and cons of both qualitative and quantitative relationships in their descriptive power. Teachers can support student reflection on their performance by presenting a phenomenon to students and ask them to list questions that can be answered qualitatively and quantitatively.

Teachers can explicitly support students in *distinguishing between scientific and non-scientific questions* by presenting students a list of questions, some that are scientific (address the natural world and can be tested empirically) and some that are not. In small groups or as a class, group the types of questions into a T-chart of scientific and not scientific questions. Teachers can reflectively support the practice of distinguishing between scientific and non-scientific questions by having students examine the T-chart that was created. Then have students make a list of the qualities of scientific questions and the qualities of non-scientific questions.

4.1.1.3 Explicitly and Reflectively Supporting Process Goals for Asking Questions and Defining Solutions for Grades 6–8

The practice of asking questions and defining problems for students in grades 6–8 builds on grades K–5 experiences and progresses to specifying relationships between variables, clarify arguments and models. A breakdown of the learning tasks that can serve as process goals for students in grades 6–8 are:

- Determine relationships between independent and dependent variables and relationships in models
- Ask questions that require sufficient and appropriate empirical evidence to answer
- Ask questions that challenge the premise(s) of an argument or the interpretation of a data set
- Define a design problem that takes into consideration scientific knowledge that may limit possible solutions

Teachers can explicitly support students in *determining relationships between independent and dependent variables and relationships in models* by drawing a model of the phenomenon of interest for the class and guiding students to identify the variables in the phenomenon. Ask students to form groups to identify relationships between the variables and form a question that can be tested. Teachers can support students in reflecting on their performance of determining relationships between independent and dependent variables by presenting the questions that students from the whole class formed earlier. Ask students to identify the independent and dependent variables from each question.

Teachers can explicitly support students in the process goal of *asking questions that require sufficient and appropriate empirical evidence to answer* by asking students how they might measure the variables so that they minimize bias from the questions the students asked about the phenomenon of interest. The teacher can support student reflection on their performance of the process goal by asking students to explain if the questions require empirical evidence and the rationale behind their response.

Teachers can explicitly support students in *asking questions that challenge the premise(s) of an argument or the interpretation of a data set* by giving students a research question, companion data set, and interpretation. Point out to students the relationships among the research question, the data collected, and the interpretation and how you identified them. Teachers can support reflectively learning about asking questions that challenge the premise(s) of an argument or the interpretation of a data set by having students write down questions that they wonder about the interpretation of the data set.

Teachers can explicitly support students in grades 6–8 in *defining a design problem that takes into consideration scientific knowledge that may limit possible solutions* by presenting completed engineering design that is used in society. As a whole class, identify the scientific principles that are related to the design. Teachers can support student reflection on their ability to meet the process goal by presenting an engineering design problem to students and asking them to identify the scientific principles that are related to the problem and list how they might cause constraints (e.g. gravitational force could constrain the height of a tower).

4.1.1.4 *Explicitly and Reflectively Supporting Process Goals for Asking Questions and Defining Solutions for Grades 9–12*

Students in grades 9 through 12 should be able to perform key learning tasks about asking questions and defining problems by building on

Asking Questions and Defining Problems   53

K through 8 experiences and progress to formulating, refining, and evaluating empirically testable questions and design problems using models and simulations. The breakdown of key learning tasks that can serve as process goals for this practice are:

- Ask questions that arise from unexpected results, to clarify and/or seek additional information
- Evaluate a question to determine if it is testable and relevant
- Ask and/or evaluate questions that challenge the suitability of the design
- Define a design problem that involves constraints that may include social, technical, and/or environmental considerations.

Teachers can explicitly support students in *asking questions that arise from unexpected results, to clarify, and/or seek additional information* by presenting a research question, data and results of an investigation that had major flaws. As a whole class, have students ask questions about the investigation and results as the teacher guides students to find difficulties with the investigation. Teachers can help students reflect on their performance by presenting the class with a research question, companion data, and vague or counter-intuitive results. Have students work collaboratively to ask questions to clarify and/or seek additional information from the results.

Teachers can explicitly support *student evaluation of a question to determine if it is testable and relevant* by reminding students what it means for a question to be testable and relevant. Have students identify a phenomenon they have studied in the past and have them ask appropriate questions about the phenomenon. Teachers can help students be reflective about their ability to perform the practice by providing a list of questions, some that are testable and relevant and some that are not, to students. Ask students to evaluate each question for testability and relevance.

Teachers can explicitly support students in *asking and/or evaluating questions that challenge the suitability of the design* by presenting a design challenge to students and have them brainstorm questions about the design. Then ask students to group them into suitable and not suitable for the design. Peer review their reasoning. Teachers can support student reflection on the practice by providing students a design challenge with a list of questions that are both suitable and not suitable. Have the students identify the unsuitable ones and give suggestions for how to adapt these questions to make them more suitable for the design challenge.

Teachers can explicitly support student practices of *defining a design problem that involves constraints that may include social, technical, and/or environmental considerations* by presenting students with a design for an engineered tool or system. Guide them to think about the impact on society, technology, and/or the environment for this tool or system. Teachers can help students reflect on their performance of the practice by reviewing a design challenge with the class. Then have students begin with physical constraints and then extend their list of constraints to include social, technical, and/or environmental considerations.

4.2 Strategies for Teachers to Support Student Practices

To further illustrate the practice of asking questions and defining problems, this section will focus on the SRL processes of *self-efficacy* in the Forethought phase, *attention focusing* in the Performance phase, and *self-evaluation* in the Self-Reflection phase as seen in Figure 4.1. The focus will be on these processes because they are most illustrative for the practices of asking question and defining problems. In other chapters discussing other practices, the book will focus on other SRL processes so that the whole cycle will be discussed across the chapters in this book.

4.3 Instructing Students in Self-Regulating Their Learning about Asking Questions and Defining Problems

The Metacognitive Promoting Intervention-Science (MPI-S) is a coaching strategy for supporting student SRL processes and is based on a research-

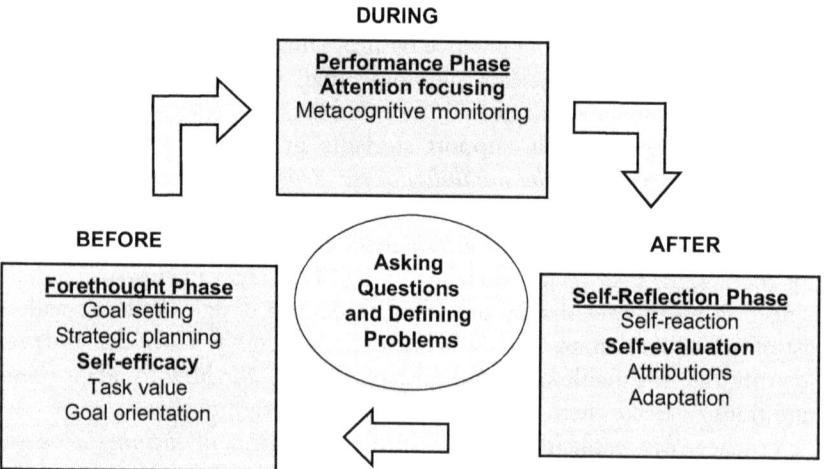

Figure 4.1 SRL processes for asking questions and defining problems

backed teaching strategy (Peters, 2009; Peters & Kitsantas, 2010; Peters-Burton & Burton, 2020). MPI-S helps students become aware of their own SRL because, for example, it prompts students about sources of self-efficacy by explicitly demonstrating the practice which gives students who do not have experience with the practice an entry point to learning. MPI-S also assists students in focusing their attention on the key features of asking questions and defining problems, and asks students to reflect on their successes and failures in their performance of the practice. This coaching strategy works well for lessons involving science and engineering practices because it engages students with the processes and approaches to thinking scientifically in a tangible and systematic way.

MPI-S is a suite of curricular tools, made up of a suite of checklists and questions that can be incorporated into established lesson plans to support student SRL strategies. The implementation of MPI-S consists of four steps: Modeling, Emulation, Self-Control, and Self-Regulation. The steps of MPI-S are the same ones as the coaching strategy founded by Zimmerman (2000). In the modeling phase, the teacher demonstrates key features of asking questions and defining problems. The student considers their own forethought processes for the skills that the teacher models. In the emulation phase, the teacher provides a checklist of key features of asking questions and defining problems. The students use the checklist as a tool for forethought and sometimes the performance processes. In the self-control phase, the teacher provides a short checklist and asks the student questions about their learning strategies. The students monitor their progress for asking questions and defining problems using the checklist and the questions. In the self-regulation phase, after students obtain feedback, the teacher asks students to explain how and why they used key features of asking questions and defining problems. The students identify the instances of learning and assess their quality of asking questions and defining problems. If a teacher notices that students are not able to perform the learning tasks in a particular phase, they can move to the prior phase until the student masters that phase.

Using this coaching approach, the teacher initially supports students explicitly through modeling and then drops the level of support so that students are able to articulate how they understand how to ask questions and define problems (or any of the science and engineering practices) independently. The first two steps of MPI-S (modeling and emulation) are instructional and the second two steps (self-control and self-regulation) are an assessment designed to pinpoint student learning to inform instruction. MPI-S does not increase the time it takes to teach science and engineering practices, because the teaching focuses on student use of science and engineering practices that are tangible at the same time, and teachers create opportunities for students to be more aware of their learning strategies.

4.3.1 Modeling

Modeling is the first step in the MPI-S teaching strategy and is aligned to forethought processes of SRL, in this case self-efficacy (Figure 4.1). Students are often underexposed to the ways scientists think and conduct their work (Hogan, 2000). Therefore, they may not have high self-efficacy to accomplish the practices of asking questions and defining problems. Giving students support to improve their self-efficacy for asking questions and defining problems can help them take more academic risks, which can result in student persistence in the face of failure. High self-efficacy also helps students to challenge themselves and stretch their perceived ability to meet goals that improve the way they perform a science and engineering practice. On the other hand, low self-efficacy can cause students to avoid mastery because they are either afraid to try or are not willing to persist in their learning. According to Bandura (2002), there are four sources of self-efficacy, which are listed from most influential to least influential:

- having performed a similar task and have been successful
- having witnessed a peer or someone they trust perform a similar task successfully
- being encouraged by a mentor
- being excited (or conversely being anxious) about taking on the task

The modeling step in MPI-S helps students with their forethought processes such as self-efficacy because it asks students to articulate their learning goal, and helps them evaluate their self-efficacy and value in the task. Students may realize through teacher modeling that they have performed a similar task and been successful or by witnessing the modeling, resulting in their ability to visualize their success. Alternatively, students may witness the modeling and become overwhelmed. Therefore, teachers should take care to break down the key features of the practice of asking questions and defining problems so that students can set smaller goals and have higher self-efficacy beliefs. Teachers can also add concrete forms of encouragement for students doing the practice during later phases of the coaching strategy.

Modeling is much like a cognitive apprenticeship (Collins, Brown, & Newman, 1989) where the mentor (teacher) does the activities in full view of the apprentice (students), but at the same time talks aloud about rationale, choices, and decision points with the intention that the apprentice will be able to adopt the same practices. The role of the student in this first step of MPI-S is to notice key features of the skill as demonstrated by the teacher and ascertain the overall sense of the outcome. Students will learn how to reach the outcome through understanding process goals in later steps of the teaching strategy.

For example, the teacher may design an investigation that involves an engineering design process so that students begin to master defining problems in grades 3 through 5. Consider an engineering challenge that uses potato chip packaging technology to demonstrate that gas takes up space, just like solids or liquids. Using household items, students define problems with potato chip packaging technology, conduct investigations, and use an engineering design process to make sense of the idea that gas takes up space (has volume).

Recall from the list of key features of asking questions and designing problems that students in the grade bands 3 to 5 are expected to master:

- Asks questions about what would happen if one variable would change
- Distinguish between qualitative and quantitative relationships
- Distinguish between scientific and non-scientific questions

In the modeling step of the coaching model, teachers can bring in different ways potato chips are stored as a demonstration. Teachers can explain how potato chips are transported and that the goals of the packaging are to keep the chip from becoming stale and from being broken. Teachers can guide students to brainstorm what they wonder about potato chip packaging such as why they do not fill the containers all of the way to the top, and how a soft bag can protect chips from being broken. Teachers can also help students think about how a can or bag releases gases when they are first opened. Teachers can ask students to identify variables that may be related to potato chip packaging technology and then point out when one changes how to observe the others. When students identify variables, the teacher can model for students how they might connect the change of one variable to another and how to measure it.

Teachers can then also model the difference between qualitative and quantitative relationships when variable relationships are identified. As a class, the teacher can guide students to group the types of questions into a T-chart of testable and not testable questions, pointing out the reasons why questions are testable or not. Most importantly during this step in the coaching phase, teachers should talk out loud about their decisions and thinking process. Student can then recognize the types of thinking that engineers use when defining problems.

Reflection question for teachers: What are the ways in which you can model asking questions and defining problems in an investigation in your classroom? Refer to the list of key features of asking questions and defining problems to explicitly model process goals that are developmentally appropriate for your students.

4.3.2 Emulation

It is during this second step of the coaching strategy when the shift from teacher-led to student-led activities begins. The emulation step is related to the SRL phase of forethought (Figure 4.1), like modeling, but is different because it guides students to set their own goals for learning about asking questions and defining problems. During emulation, the role of the student is to replicate the science and engineering practices that the teacher models when they are given a similar task as the model. However, the student does this with considerable support, and the teacher provides the students with a checklist to set goals, consider their self-efficacy, task value, and goal orientation on a science and engineering skill related to asking questions and defining problems in the investigation. The checklist should begin with the outcome goal for the design challenge and then list the process goals that will help students define the problem they are going to pursue in the investigation. For example, when helping students to define a problem for the potato chip packaging technology, students are supported with a checklist of the following statements that have been addressed in the modeling step of the coaching strategy:

- I have identified the goals for potato chip packaging (outcome goal or design challenge)
- I have identified different ways potato chip packaging can change (variables)
- I can list at least three things that can be changed on the design of potato chip packaging
- From the list of three things that can be changed on the design of the packaging, I can identify qualitative and quantitative relationships
- I can list three questions to ask about improving the design of potato chip packaging
- I can list the constraints of the design
- From the list, I can identify what questions can be tested and what questions cannot be tested

- If I think I cannot do any of the things on the checklist, I will look for help from a peer or the teacher

Students use these statements to help them reinforce their self-efficacy for asking questions and defining problems. Like cognitive apprenticeships, MPI-S helps students who may not have had prior access to ways of knowing in science and engineering by explicitly pointing out not just how to do the practice (procedural knowledge), but also why they are doing the practice (epistemic knowledge). In later lessons, teachers can use different checklists for other science and engineering practices.

To begin making a checklist for your students, list some metacognitive prompts for an investigation that you do in the class related to asking questions and defining problems. Refer to the list of learning tasks for each developmental stage for the practice of asking questions and defining solutions. To begin, identify the key characteristics of the practice that you would like your students to master, and then create a list of two to three prompts that help students monitor if they have made progress to identify the key characteristic from the skill.

4.4 Assessing Students in Self-Regulating Their Learning about Asking Questions and Defining Problems

4.4.1 Self-Control

The self-control step of MPI-S is related to the performance phase of SRL (Figure 4.1) because it helps students monitor their performance and focus their attention on learning about science and engineering practices. Students engaged in MPI-S to this point have observed what they are supposed to be accomplishing through the teacher model (Modeling) and have attempted similar skills and knowledge with support from the teacher (Emulation). In the third step, the teacher continues to support student self-regulation of learning about science and engineering practices but reduces support to allow students to actively reflect on their metacognitive strategies. The third and fourth steps of this coaching approach can also be used to assess how well students are beginning to self-regulate their new skills. Teachers should provide students with a more difficult attempt at the skill they are trying to build and give students only a few basic standards from which to check. Students are expected to take over more responsibility for learning and the teacher acts as the facilitator by providing basic support, only intervening when misconceptions arise. Using the example of the potato chip packaging design challenge for defining problems, teachers support students by providing the following shortened checklist during their investigation:

- I can list three questions to ask about improving the design of potato chip packaging
- I can list the constraints of the design
- From the list, I can identify what questions can be tested and what questions cannot be tested

To check for appropriate student metacognition in this step, teachers should ask a few questions about the choices that students make when they perform the practice such as

- What are the variables that you want to test for your design challenge?
- How do you know the difference between a qualitative and a quantitative measurement?
- How do you know that the problem you have defined will meet the design challenge?

If students struggle to answer these questions, then they likely have not mastered the learning tasks of the practice. Teachers can help students by returning to the modeling phase and explicitly teaching the difficult practices, then giving students another chance to emulate. When students can answer these questions in a way that demonstrates their mastery of defining problems, then teachers can fade all support in the next step, self-reflection.

To create student supports for the emulation phase of coaching SRL, list your shortened checklist of metacognitive prompts for an investigation that you do in the class related to asking questions and defining problems. Follow that up with some questions for the students to answer about the practice. Use the example discussed in this section as a model.

4.4.2 Self-Reflection

In the self-reflection step of MPI-S, students perform the targeted practice entirely on their own and reflect on the outcome. This last step of MPI-S is aligned to the self-reflection processes of SRL and helps students to self-evaluate their performance and attribute their successes and failures to sources of learning. The self-reflection step builds upon the self-control step because students are expected to regulate their learning without any support. Students should be able to demonstrate they can both understand and implement the science and engineering practice (asking questions and

Asking Questions and Defining Problems

defining problems) without any teacher support. In this step, the teacher gives the learning task and ensures that the student was able to accomplish it in a way that parallels the science and engineering practice. The teacher can decide if the student has mastered the science and engineering practice by evaluating student answers to the questions about rationale for their learning processes. Questions that a teacher could ask regarding asking questions and defining problems for grades 3 through 5 for the potato chip packaging design challenge are:

- How confident were you in defining problems when you started this challenge? When you ended this challenge? Explain why (self-efficacy)
- Explain how you identified variables of the potato chip packaging (attention focusing)
- Give an example of a quantitative problem and a qualitative problem that you identified in potato chip packaging (attention focusing)
- How did you use the constraints to define the problem you will investigate? (Self-evaluation)
- How did you know that your question or problem was testable? (Self-evaluation)
- What about asking questions and defining problems you think you do well? Why?
- What about asking questions and defining problems you think you still need to practice? Why?

To create your own questions, list some ways that a student can show you that they are self-evaluating in a productive way for the investigation. Then list questions that will address student self-evaluation for an investigation in your science class. In the next section, there will be two case studies to demonstrate how different processes of self-regulated learning could help or hinder students in their learning of asking questions and defining problems.

4.5 Teacher–Student Cases

4.5.1 Case Study: Adaptive

The case study featured in this chapter as an example of an adaptive and effective learner for defining problems will focus on:

- Self-efficacy in the forethought phase
- Attention focusing in the performance phase
- Self-evaluation in the self-reflection phase

Clarissa, a 5th grade student, is working with her group to define the problem for the following challenge: *What can you change about potato chip packaging to make it more ecologically friendly while still keeping the chip fresh and unbroken while being transported?* At first, Clarissa was overwhelmed by defining the problem because she did not have much experience with potato chip packaging (lack of self-efficacy). However, one of her groupmates told her to make a model of a chip bag with a re-sealable sandwich bag. When Clarissa filled the bag with air and quickly sealed it, she could see how the bag served as a model for the packaging. She then felt as though she could work through the definition of the problem with her group (higher self-efficacy).

Clarissa knew from the checklists that she should know the criteria and the constraints of the challenge in order to define the problem. She was able to find the goal for the challenge from the teacher directions and used the checklist the teacher gave her to focus her attention on being able to identify variables and potential relationships in potato chip bags (attention focusing). She manipulated the re-sealable bag as a model to figure out what the variables were, and then added a list of ecologically friendly materials that could be switched out for the plastic bag.

When Clarissa and her group finalized the list of criteria and constraints, they discussed the design they could pursue to make a more ecologically friendly potato chip package. When Clarissa felt like part of the design was not acceptable, she went back to the list of criteria and constraints to see if the design fits. In this way, she was able to evaluate the problem that her group defined (self-evaluation).

Reflection question for teachers: Reflect on the adaptive case study and think about the following question: What student supports for defining problems can you use in your classroom from this case study?

4.5.2 Case Study: Maladaptive

The case study featured in this chapter as an example of an ineffective learner that needs support for asking questions and defining problems will focus on:

- Self-efficacy in the forethought phase
- Attention focusing in the performance phase
- Self-evaluation in the self-reflection phase

Bridget, a 5th grade student in Clarissa's class, liked to build things out of cardboard boxes, but she had never heard of the field of engineering. As

Bridget was joining her new group for the potato chip packaging design challenge, she was not feeling confident in her ability to do well with this project (low self-efficacy). She did not have a foundational idea of what she was about to do. As the group got started with defining criteria and constraints, she zoned out because she did not want to look silly in front of her new group. When the teacher noticed that Bridget was not contributing, the teacher asked her to give one idea. Bridget just said she was not sure what was most important and her ideas were already said (attention focusing). When Bridget's group finalized the problem that they were going to investigate, replacing bags with boxes, she wanted to contribute, but she was not sure of the variables that were being investigated (self-evaluation). In her mind, her continued confusion was justified because she expected that she was not going to do well anyway.

4.5.3 Supports the Teacher Can Provide for Bridget in This Case Study

4.5.3.1 Self-Efficacy

Bridget's low confidence for the design challenge prevents her from stretching academically and taking risks in learning. The teacher could help Bridget see that her experience with building things from cardboard boxes would help her in any design challenge and the decisions she makes in building apply to school projects, particularly because her group wanted to build a box as the container. The teacher could talk to Bridget about how she approaches building with cardboard and connect her activities to engineering. For this project, the teacher could offer chunking of the tasks to help Bridget feel less overwhelmed. The teacher and Bridget should work together to decompose the large task and the teacher can talk out loud about their thinking process to show Bridget how defining problems is done. This would help with two of the more influential sources of self-efficacy, prior experiences and observing someone else doing the activity.

4.5.3.2 Attention Focusing

Bridget was not able to wrap her head around the outcome goal and was not able to break down the process goals, even though she had some experience with it in an out-of-school setting. Providing a checklist of the learning tasks could help Bridget, but only if she can see what they mean in terms of her own experiences. Bridget could also document when her tasks get accomplished, so that she continues to boost her self-efficacy.

4.5.3.3 *Self-Evaluation*

It is not surprising that Bridget had a difficult time evaluating her own performance because even by the end of the activity of defining the problem, she was unaware of the learning tasks. She was therefore unable to compare her performance against the expected standard. By speaking with Bridget about any experiences she might have had outside of school related to creating products such as the packaging, the teacher might be able to help Bridget see what the process goals are for defining a problem. This may, in turn, help Bridget build higher self-efficacy which can help her contribute more in the next design challenge.

Reflection question for teachers: Reflect on the maladaptive case study you just read and answer the following question: What student supports for asking questions can you use in your classroom from this case study?

4.6 Designing Lessons That Use the Practice of Asking Questions and Defining Problems

In order to plan for teacher modeling of the practice of asking questions and defining problems, teachers can review a lesson plan for ways they can model the key learning tasks in asking questions and defining problems in an investigation. Teachers can look for opportunities in their lesson to point out to students:

- Identification of variables in a phenomenon
- Relationships in a phenomenon
- When questions are testable and scientific
- Ways to identify criteria and constraints in a problem

Teachers can then make checklists for students by listing the key features of doing the science and engineering practice, and then convert them into student-friendly language for a checklist bullet. For example, a key feature from asking questions and defining problems is to define a design problem that takes into consideration scientific knowledge that may limit possible solutions. The checklist bullet item for this key learning task can be "When I define a design problem, I go back to see if it fits into the laws of the natural world." Teachers should make at least one bulleted item for each key feature of the practice. Making more than one item could potentially help students by understanding the practice from different perspectives.

4.7 Teacher Reflection on Implementation of Lesson Featuring Asking Questions and Defining Problems

Being a reflective practitioner is important for supporting student self-regulated learning. Teachers should consider the following questions for the practice of asking questions and defining problems:

- What worked well in supporting students asking questions and defining problems?
- What did not work well in supporting students asking questions and defining problems?
- How will I change this the next time I teach students to ask questions and define problems?

CHAPTER 5

Developing and Using Models

Chapters 4 through 11 in this part have a parallel design. Each of the chapters deals with a particular science and engineering practice, and this chapter focuses on developing and using models. In each chapter, the practice is dissected into distinct and clear learning tasks that serve as process goals for learning the practice. These tasks are then examined within the context of a self-regulated learning cycle and coaching strategies for instruction and assessment are emphasized. The instruction and assessment strategies are contextualized for students in grades 9–12 and focus on developing and refining a model for electrolysis. The tasks are reassembled into two case studies – one positive and one negative – to demonstrate how the learning tasks can be used by students and how teachers can support students learning how to develop and use models.

5.1 Learning Tasks in the Practice

The Next Generation Science Standards (NGSS Lead States, 2013) explain that models are used in both science and engineering to represent ideas for the purposes of explanatory power. Models can include tools such as diagrams, drawings, physical replicas, mathematical representations, analogies, and computer simulations. In the classroom, models can be developed by students to demonstrate their prior and recent knowledge about the natural world, a system, or a design. As students learn more about a topic, they can adapt and improve the model to demonstrate their learning. Models can be used in combination with other science and engineering practices to lay out background knowledge such as to ask questions and define problems, to analyze data, and to communicate explanations.

These overall definitions of developing and using models have been further broken down into grade bands by NGSS of K-2 (aged 4 to 7), 3–5 (aged 8 to 11), 6–8 (aged 12 to 14), and 9–12 (aged 15 to 18) as seen later in the chapter. The learning task breakdown explains each new objective for the science and engineering practice of developing and using models for the grade bands. The breakdown begins with the learning tasks from K-2 and builds on the new practices that are to be learned in the successive grade band (NGSS@NSTA, 2014). From there, the learning tasks involved in each objective for the practice are explained, and these learning tasks represent what students should be able to do to master the practice. The explanation of the task breakdown also provides examples of how a teacher can support student process goals in an explicit way and in a reflective way. Educational research on epistemic knowledge in science has shown that students are less likely to learn an aspect of scientific thinking implicitly. However, it is more likely that students will learn how to think like a scientist when taught explicitly by the teacher and given intentional opportunities to reflect on how the aspect was demonstrated (Khishfe & Abd-El-Khalick, 2002; Peters, 2012; Peters & Kitsantas, 2010; Peters-Burton, 2015). This literature is used to offer practical ways to enhance student learning for science and engineering practices in this part of the book.

Breaking the practice down into process goals is not only important for helping students who may not have had exposure to doing science and engineering, but it is also important to situate the learning task in a cycle of self-regulated learning. In order for self-regulated learning to be productive, one must begin with a well-defined learning task. As explained in Chapter 3, goal setting is key to being successful with self-regulated learning, and the goal should be discrete and composed of smaller, more proximal process goals that lead up to the outcome goal. In the overall definition of this practice, teachers can identify key characteristics of developing and using models such as using a visual tool to describe baseline knowledge of students. Students can then have experiences in other practices that can influence the explanation by the visual tool. Students can revise, add detail, or create new information on the model to increase the explanatory power of the visual model. One thing to keep in mind with models is that they rarely describe the phenomenon, system, or design in total and are mainly used to emphasize a small set of information about the phenomenon, system, or design. Thus, students have an opportunity to present the pros and cons of their own models and models from other sources when involved in this practice. The key characteristics listed here for developing and using models can serve as outcomes goals for learning the skills. However, recall that in order for students to master the practices, they also need process goals that consist of smaller steps used for reaching the outcome goals. The learning tasks for developing and using models that can serve as the foundation for process goals for students can be found in the next section.

5.1.1 Process Goals for Developing and Using Models

In grades K-2, the NGSS communicate the expectation for students that modeling in grades K through 2 builds on prior experiences and progresses to include using and developing models (i.e., diagram, drawing, physical replica, diorama, dramatization, or storyboard) that represent concrete events or design solutions (NGSS@NSTA, 2014). Student learning tasks for that expectation are broken down into:

- Recall and organize prior experiences related to the phenomenon
- Distinguish between a model and the actual phenomenon, system, or design.
- Draw a model from baseline prior experiences
- Revise the model based on observations or measurements from investigation experiences

5.1.1.1 Explicitly and Reflectively Supporting Process Goals for Developing and Using Models

Research in education has demonstrated that students can learn about science more effectively when teachers explicitly and reflectively support learning tasks (Peters-Burton, 2017). Explicitly teaching a practice does not mean lecturing to students about the practice. Instead, teachers can explicitly teach the behaviors that scientists and engineers value by modeling the practice of developing and using models. When reflectively teaching the practice, teachers give students a chance to independently show what they know about performing the practice. Teachers can determine if they need to reteach the practice or give students another chance to emulate the practice when they assess the student's reflection on the practice.

Teachers can explicitly support students in *recalling and organizing prior experiences related to the phenomenon* by engaging students with the general phenomenon for which they will be asking questions or defining solutions. Then, provide a graphic organizer to students to record their prior experiences. From this record, ask students to share experiences and as a whole class, develop a webbed concept map, drawing, or physical replica that incorporates student experiences in the model. When explicitly teaching this practice, use different types of models so students see a variety of examples. Teachers can support student reflection on their learning about the practice by giving students a copy of the model that the teacher and class developed. Have students individually highlight what they already knew in one color. In another color have them highlight features of the model that they learned about and now know because of the class discussion. In a third color, have students highlight features of the model that they are still confused about or parts of the model that were not discussed or represented in the model.

Teachers can explicitly teach students to *distinguish between a model and the actual phenomenon, system, or design* by showing an example of a phenomenon, system, or design to students as a whole class. Ask students to develop a model for the phenomenon, system, or design in small groups. Have the groups explain how the model represents, but is different from the phenomenon, system, or design. Guide students when they have difficulty explaining the differences. Teachers can support reflection for students by providing a collection of examples of phenomenon, systems, or designs, along with models representing them. Ask students to distinguish between the model and the phenomenon, system, or design and explain their thinking.

Teachers can explicitly teach students to *draw a model from baseline prior experiences by selecting a phenomenon, system, or design*. Demonstrate how the phenomenon has variables that have amounts, relationships, relative scales (bigger, smaller), and patterns. As a class, make decisions about what amounts, relationships, scales, and patterns to represent as you construct the model. Teachers can reflectively support students by providing students with a phenomenon, system, or design and a graphic organizer that has students list variables, relationships, and other aspects that the model should show. Ask students to use their completed graphic organizer to develop a web concept map, labeled diagram with micro and macro details, or a flowchart as appropriate to the assignment.

To explicitly support students to *revise a model based on observations or measurements from investigation experiences*, teachers can collaboratively develop a model with the class. Before an investigation, develop a model of what they are about to study by guiding student thinking as a class. Once the investigation is complete, again collaboratively revisit the model with the whole class and make changes in another color to demonstrate what they learned about the phenomenon, system, or design from the investigation. Things that are still murky can be labeled for clarification when performing other practices such as analyzing and interpreting data.

To support students reflectively, teachers can have students develop a model of what they are about to study. Once the investigation is complete, have the students revisit the model and make changes in another color to demonstrate what they learned about the phenomenon, system, or design from the investigation.

5.1.1.2 Explicitly and Reflectively Supporting Process Goals for Developing and Using Models for Grades 3–5

The practice of developing and using models for students in grades 3 through 5 builds on student experiences during K through 2 and progresses to building and revising simple models and using models to represent events and design solutions. A breakdown of student learning tasks related to this practice are:

- Evaluate the pros and cons of a given model
- Collaboratively develop a model that illustrates both variables and relationships between the variables
- Expand the development and use of models by using an analogy or abstract idea to represent a phenomenon, system, or design
- Make predictions using the model
- Use the model to explain cause-and-effect relationships of a system (natural or designed)

Teachers can explicitly teach students *to evaluate the pros and cons of a given model*. After an investigation, examine the revised model with the whole class. Give a few examples of how the model is helpful and how the model hides some aspects of the phenomenon. Ask students to find other helpful and hidden aspects of the phenomenon, system, or design based on the model. Teachers can support students' reflection on evaluating the pros and cons of a given model so that students can monitor their progress. When studying another phenomenon, system, or design, ask students to complete a T-chart of pros and cons of the model. The reason for doing this after an investigation is that students have the most complete knowledge about the phenomenon, system, or design at that time.

Teachers can help students *collaboratively develop a model that illustrates both variables and relationships between the variables* by being explicit and talking aloud about their thoughts while performing the practice. As a whole class, introduce students to a phenomenon, system, or design. Have students identify variables for the model. Once the students have identified the variables, show how two variables are related. Ask students to find other relationships. Teachers can then help students to reflect and assess their progress on learning the practice. Have students investigate a phenomenon, system, or design and in small groups list the variables. Have small groups collaboratively develop a model of the phenomenon, system, or design using the list of variables and identifying any relationships between variables.

Teachers can support student learning of *expanding the development and use of models by using an analogy or abstract idea to represent a phenomenon, system, or design.* They can model the practice explicitly by selecting a phenomenon, system, or design. Using the actual phenomenon or a model representing the phenomenon, demonstrate how an analogy fits the model or phenomenon. Teachers can help students to reflect on their skill of expanding the development and use of models by providing students a phenomenon and have them conduct an investigation on it to understand the relationships and variables. Then ask students to find an analogy that fits the phenomenon and explain their reasoning for the fit.

Teachers can be explicit *about making predictions using the model* to show students how scientists would approach using this skill. Before beginning an investigation, collaboratively develop a model with the whole class on the phenomenon, system, or design. Refer to the research question or defined problem and make a prediction related to the question or problem from the design. Talk aloud to the students about how your decisions are being made as you make predictions from the model. Point out what factors you are considering to make the prediction. Teachers can help students reflect on their ability to make predictions using the model. Before beginning an investigation, have students create a model of the phenomenon, system, or design. Ask students to make predictions and explain their decision-making process for that prediction using the model.

Teachers can demonstrate explicitly to students how to *use the model to explain cause-and-effect relationships of a system.* Collaboratively develop a model with the whole class on the phenomenon, system, or design. Refer to the research question or defined problem and identify the independent variable and the dependent variable. Be explicit with the students about how your decisions are being made as you identify the independent and dependent variables from the model. Point out the cause-and-effect relationship from the model. To help students be reflective about their newly learned skill, have them revise the models of the phenomenon after students complete an investigation. Ask students to identify independent and dependent variables in the model and explain the cause-and-effect relationship using the model.

5.1.1.3 *Explicitly and Reflectively Supporting Process Goals for Developing and Using Models for Grades 6–8*

The practice of developing and using models in grades 6 through 8 builds on students experiences from grades K through 5 and progresses to developing, using, and revising models to describe, test, and predict more abstract phenomena and design systems. The following learning tasks can be used as process goals for students learning this practice:

- Determine relationships between independent and dependent variables and relationships in models
- Demonstrate changes in a model if one of the variables is changed
- Develop a model illustrating both what is known and what is uncertain
- Develop a model for a phenomenon, system, or design that cannot be directly observed
- Use a model for a phenomenon, system, or design that cannot be directly observed in order to generate data from the model

Teachers can explicitly teach students how to *determine relationships between dependent and independent variables in models* so that students can understand what they need to master. As a class, draw a model of the phenomenon of interest and have students identify the variables in the phenomenon. Ask students to form groups to identify relationships between the variables and form a question that can be tested. To help students be reflective about their newly learned skill, present the questions that students developed from the whole class discussion (during the explicit teaching of the relationships and variables in a phenomenon). Ask students to identify the independent and dependent variables from each question and explain their reasoning.

Teachers can explicitly help students understand how to *demonstrate changes in a model if one of the variables is changed*. As a class, draw a model of the phenomenon of interest and have students identify the variables in the phenomenon. Ask students to form groups to explain how when one variable changes, the outcomes changes. Discuss group product in a class discussion. Teachers can support student reflection on their skill in determining changes in a model if one variable is changed by presenting a model and one variable that can change in a phenomenon. Ask students to identify that when each variable changes, the others change as well with their reasoning.

Teachers can be explicit about *developing a model that illustrates what is known and what is uncertain on the model*. Show students a model and talk aloud about your thinking while you highlight a few things in one color that you are sure about regarding the variables and relationships on the model. In another color highlight features of the model that were not discussed or represented in the model, revealing your thinking by talking aloud about your reasoning. Then to help students be reflective, teachers can have students develop a model from a phenomenon, system, or design. When students begin an investigation, have them individually highlight what they already knew in one color. During or after the investigation, in another color have them highlight features of the model that they learned about from doing the investigation. In a third color, have students highlight features of the model that they are still confused about or parts of the model that were not discussed or represented.

To explicitly teach students how to *develop a model for a phenomenon, system, or design that cannot be directly observed*, teachers can select a phenomenon, system, or design that cannot be directly observed. Collaboratively as a class develop a model while discussing aloud your thinking as you put the model together. Teachers can also help students to reflect on how well they are accomplishing the task by having students in small groups identify a phenomenon, system, or design that cannot be directly observed. Have them collaboratively develop a model and explain the variables and relationships in the model.

Teachers can teach students how to *use a model for a phenomenon, system, or design that cannot be directly observed in order to generate data from the model* by first selecting such a phenomenon, system, or design that cannot be directly observed. Collaboratively as a class develop a model while discussing aloud your thinking as you put the model together. Show how putting inputs into the model generates data as outputs. Then teachers can help students reflect on their proficiency in the skill by having students in small groups identify a phenomenon, system, or design that cannot be directly observed. Have them collaboratively develop a model and explain the variables and relationships in the model. Have them collect outputs of the model after they input the independent variables and peer review the outcomes.

5.1.1.4 Explicitly and Reflectively Supporting Process Goals for Developing and Using Models for Grades 9–12
Students in grades 9 through 12 should be able to perform key learning tasks about developing and using models by building on grade K through 8 experiences and progress to using, synthesizing, and developing models to predict and show relationships among variables between systems and

their components in the natural and designed world(s). The following learning tasks can serve as process goals for students in grades 9–12:

- Evaluate a model for strengths and weaknesses and revise the weaknesses of the model
- Generate different types of models given different phenomenon, systems, or designs
- Use a model to identify variables, relationships, and mechanisms in a phenomenon, system, or design
- Develop an abstract model that can produce outputs given inputs

Teachers can explicitly teach students to *evaluate a model for strengths and weaknesses and revise the weaknesses of the model* so that students can understand what they need to master at this level. When students begin a unit, collaboratively develop an initial model with the class. Explain that this will be their starting point and as they learn more about the phenomenon, system, or design, they will adapt, edit, and add to the model. Make a list of strengths of the model and weaknesses of the class-developed model. To help students be reflective about their model evaluation skills, have students take the initial model developed as a class, and as they learn information about a phenomenon, system, or design, edit and/or add to the model several times throughout the unit of study. Have students explain what they changed each time they revisited the model. Have students edit the list of strengths and weaknesses of the model at the end of the investigation.

In order to be explicit to students about how to *generate different types of models given different phenomenon, systems, or designs*, a teacher can model their thinking. As a teacher progresses through the year, keep a record of the models used for different phenomenon, systems, and designs, with an eye toward introducing different types of models for different units of study. At various points in the year, point out why that type of model is useful for describing or explaining the phenomenon, system, or design. To help students be reflective about their skill in generating different models, provide students with an overview of a phenomenon, system, or design as you begin a unit. Have them select a type of model to represent the phenomenon, system, or design and explain why it is a good choice. Over the year, the students should generate different types of models.

Teachers can demonstrate how to *use a model to identify variables, relationships, and mechanisms in a phenomenon, system, or design* to show how to explicitly accomplish this skill. As a class, draw a model of the phenomenon of interest and have students identify the variables and relationships in the phenomenon. Ask students to form groups to identify mechanisms that drive the relationships and guide their thinking by visiting each group as they identify mechanisms. To help students reflect on their accomplishment in the skill, give students different models from the previous units of study. Have them identify the variables, relationships, and mechanisms in the model.

Teachers can explicitly teach students how to *develop an abstract model that can produce outputs given inputs*. First, select a phenomenon, system, or design. Collaboratively as a class develop an abstract model while discussing aloud your thinking as you put the model together. Show how putting inputs into the model generates data as outputs. Then, to help students be reflective about their new skill, have students in small groups identify a phenomenon, system, or design. Have them collaboratively develop an abstract model and explain the variables and relationships in the model. Have them collect outputs of the model after they input the independent variables.

5.2 Strategies for Teachers to Support Student Practices

For the practice of developing and using models, this section will focus on the SRL processes of *task value* in the Forethought phase, *metacognitive monitoring* in the Performance phase, and *attribution* in the Self-Reflection phase as seen in Figure 5.1. The focus will be on these processes because they are most illustrative for the practices of developing and using models. In other chapters discussing other practices, the book will focus on other SRL processes so that the whole cycle will be discussed across the book.

5.3 Instructing Students in Self-Regulating Their Learning about Developing and Using Models

A research-based coaching strategy has been developed for teachers to demonstrate strategic SRL thinking called Metacognitive Promoting Intervention-Science (MPI-S; Peters, 2009; Peters & Kitsantas, 2010). MPI-S enacts SRL because it prompts students about the value of learning a task by explicitly demonstrating the practice, which gives students who do not have experience with the practice an entry point to learning. MPI-S also assists students in metacognitive monitoring the key features of developing and using models, and asks students to reflect on the attribution of their successes and failures in their performance of the practice. This teaching strategy works well for lessons involving science and

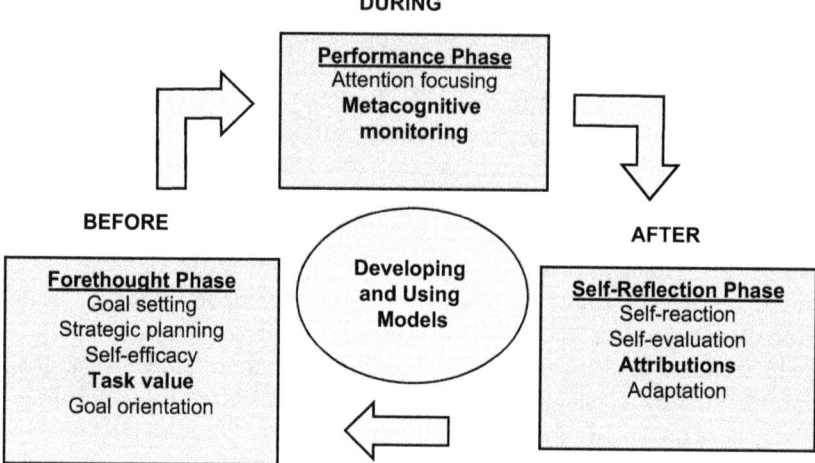

Figure 5.1 SRL processes for developing and using models

engineering practices because it engages students with the processes and approaches to thinking rationally and systematically. MPI-S is a suite of curricular tools, consisting of checklists and questions that can be incorporated into established lesson plans to support student SRL strategies.

The implementation of MPI-S consists of four steps: Modeling, Emulation, Self-Control, and Self-Regulation. The steps of MPI-S are the same ones as the coaching strategy founded by Zimmerman (2000). In the modeling phase, the teacher demonstrates key features of developing and using models. The student considers their own forethought processes. In the emulation phase, the teacher provides a checklist of key features of developing and using models for student use. The students use the checklist as a tool for their forethought processes. In the self-control phase, the teacher provides a short checklist and asks questions about student learning strategies and the students monitor their progress for developing and using models. In the self-regulation phase, the teacher asks students to explain how and why they used key features of developing and using models. Students identify the instances of learning and assess their quality of developing and using models.

Using this approach, the teacher initially supports students explicitly through modeling and then drops the level of support so that students are able to articulate how they understand how to develop and use models (or any of the science and engineering practices) independently. The first two steps of MPI-S (modeling and emulation) are instructional and the second two steps (self-control and self-regulation) assess student learning to inform instruction for teachers and to give students feedback. MPI-S does not increase the time it takes to teach science and engineering practices, because the teaching focuses on student use of science and engineering practices that are tangible. At the same time, teachers are creating opportunities for students to be more aware of their learning strategies.

5.3.1 Modeling

Modeling is the first step in the MPI-S teaching strategy and is aligned to forethought processes of SRL, in this case, task value (Figure 5.1). Students are often underexposed to the ways scientists think and conduct their work (Hogan, 2000). Therefore, they may not value learning about developing and using models. Helping students to improve their task value for developing and using models can help them persist in learning the task, even if they have low self-efficacy (Peters-Burton & Botov, 2017). High task value helps students consider the cost it takes to learn the task, such as taking time from something else they like doing more than the learning task, and consider the long-term benefits of mastering the task (Wigfield & Eccles, 1992). Seo and Taherbhai (2009) found that students who valued a certain task were more likely to find their class interesting, important, and useful.

The modeling step in MPI-S helps students with their forethought processes such as task value because it demonstrates their learning goal, and helps them evaluate their self-efficacy and value in the task. Students may realize through teacher modeling behaviors of the practice, that if they master learning this task, they can use it in other areas of their life. Therefore, teachers should take care to explain other applications of the skill of developing and using models so that students can have higher task value beliefs about the learning task. Teachers can also add other examples of valuable uses of the practice for students during later phases of the coaching strategy to enhance student task value.

Modeling is much like a cognitive apprenticeship (Collins, Brown, & Newman, 1989) where the mentor (teacher) does the activities in full view of the apprentice (students), but at the same time talks aloud about rationale, choices, and decision points with the intention that the apprentice will be able to adopt the same practices. The role of the student in this first step of MPI-S is to notice key features of the skill as demonstrated by the mentor and ascertain the overall sense of the outcome. Students will learn how to reach the outcome through understanding process goals in later steps of the teaching strategy.

For example, the teacher may design an investigation in high school (grades 9–12) that involves using a conceptual model of electrolysis to explain an abstract phenomenon in detail. The students are expected to develop a model of their ideas about electrolysis that they understand from prior experience, then observe a working electrolysis apparatus to add to their model and generate more questions. When students have gained as much information as they can from the observations, then they do research to try to answer their questions, which are usually at the molecular level.

Recall that some of the key features of developing and using models that students in high school are expected to master are:

- Evaluate a model for strengths and weaknesses and revise the weaknesses of the model
- Use a model to identify variables, relationships, and mechanisms in a phenomenon, system, or design
- Develop an abstract model that can produce outputs given inputs

In the modeling step of the coaching model, teachers can use a video or demonstration to show an electrolysis apparatus. An electrolysis apparatus can consist of a beaker with salt water, a battery with two insulated lead wires coming from each terminal, and two graphite pencils that are sharpened on either end. The pencils have graphite showing on either end to create a connection to the lead wire. See Figure 5.2 for a diagram.

First, the teacher can explain to students that the model they will develop and use is a conceptual model, so the students can begin the model with a drawing of the apparatus. After students draw the apparatus, the teacher can ask them to label the parts and check that students have labeled all appropriate equipment. Then the teacher can begin a classroom discussion of the possible variables in this system. The teacher may want to begin with macro-level variables that can be seen like the battery supplying power and then move to micro-level or molecular-level variables such as the salt dissolved in the water. Once the teacher is satisfied that all of the relevant variables are identified, students can begin the emulation level where they identify relationships and mechanisms in their model. As the discussion about variables proceeds, students may have identified some questions such as, "if the wires do not touch, what is making the closed circuit?" The teacher should keep a "parking lot" of these types of questions for students to refer to in later phases.

To modeling for task value, the teacher can discuss how in this model there are some things that are directly observable, such as the salt dissolving

Developing and Using Models 79

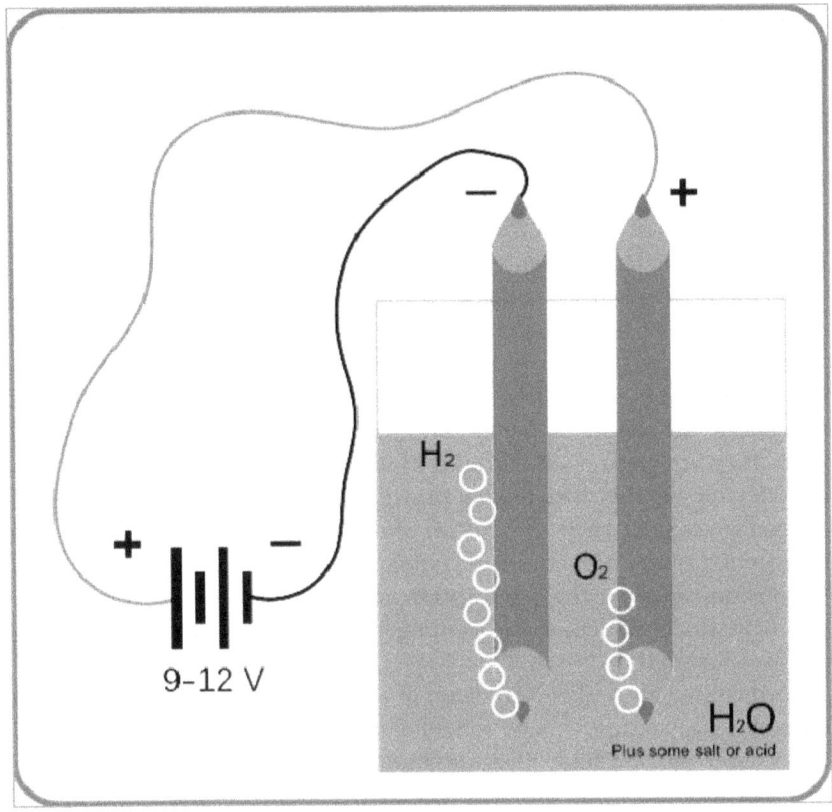

Figure 5.2 Diagram of electrolysis apparatus setup
Source: Creative Commons Wikimedia found at https://commons.wikimedia.org/wiki/File: Electrolysis.svg. © Nevit Dilmen

in the water and the bubbles forming on the pencil tips. However, there are also unobservable things such as the way the electricity is involved and what happens to the molecules in the water to form the bubbles. Being able to model all of the complex things that happen in this system is important because typically variables, relationships, and mechanisms in natural phenomenon are complex. If students only model the simple features, they are missing many of the features of the system and may have some misconceptions about relationships or mechanisms. Part of science and engineering is to thoroughly understand the context and situation so that conclusions made about variables, relationships, and mechanisms are valid, necessary, and sufficient. Teachers may help students with task value by showing them complex systems that they may encounter in their daily lives, such as workplace expectations and relationships. Fully understanding a situation can be as significant in one's own personal life as it is in science and engineering.

5.3.2 Emulation

It is during this second step of the coaching strategy when the shift from teacher-led to student-led activities begins. The emulation step is related to the SRL phase of forethought (Figure 5.1), but is different from modeling because it guides students to be the initiator of their own learning about developing and using models. During emulation, the role of the student is to replicate the science and engineering practices that the teacher modeled when they are given a similar task. However, the student does this with considerable support. The teacher provides the students with a checklist to set goals, consider their task value, and self-efficacy of the science and engineering practice in the investigation. The checklist should begin with the outcome goal for the investigation and then list the process goals that will help students define the learning they are going to pursue during the investigation.

In the emulation phase, students in small groups can build their own electrolysis apparatus, change variables, and make related observations to understand relationships. Because students are working collaboratively, they can also review the variables they have identified in the model so that they have all of the same information before they move on to explore relationships between the variables. When helping students to revise a model of electrolysis as they learn more about the system from their own manipulation of the system, students are supported with a checklist of the following statements that have been addressed in the modeling step of the coaching strategy:

- I can comprehend the goal of revising the model for electrolysis that includes variables, relationships, and mechanisms (outcome goal)
- I understand that developing more details in the model will help me make more valid conclusions about the system
- I have identified physical and abstract variables for the system
- I have manipulated all of the variables one at a time to see the results in the model
- I have had discussions with my group about any abstract variables and relationships that are not on my model

- I have revised the model based on my experiences with the electrolysis apparatus
- I have kept a record of any questions I have about the system that are not able to be answered with the apparatus

Students use these statements to help them reinforce their goal setting and task value for developing and using models. Like cognitive apprenticeships, MPI-S helps students who may not have had prior access to ways of knowing in science and engineering by explicitly pointing out not just how to do the practice (procedural knowledge), but also the rationale behind doing the practice (epistemic knowledge). In later lessons, teachers can use different checklists for different science and engineering practices.

5.4 Assessing Students in Self-Regulating Their Learning about Developing and Using Models

5.4.1 Self-Control

The self-control step of MPI-S is related to the performance phase of SRL (Figure 5.1) because it helps students metacognitively monitor their performance and focus their attention on learning about science and engineering practices. Students engaged in MPI-S to this point have observed what they are supposed to be accomplishing through the teacher model (Modeling) and have attempted similar skills and knowledge with support from the teacher (Emulation). In the third step, the teacher continues to support student self-regulation of learning about science and engineering practices but reduces support to allow students to actively reflect on their own metacognitive strategies. The third and fourth steps of this coaching approach can also be used to assess how well students are beginning to self-regulate their new skills. The overall idea for self-control is for teachers to provide students with a more difficult attempt at the skill and give students only a few basic standards from which to check. This helps students take on more responsibility for learning than in previous steps of the coaching strategy.

In this case, students will share the questions that they still have about the system with the class and attempt to build a more complete model to include mechanisms and any variables and relationships that have not yet been identified. When students cannot provide cogent answers to the questions about the system, they can do research to complete the model. Students are expected to take over more responsibility for learning and the teacher acts as the facilitator by proving basic support, only intervening when misconceptions arise. Using the example of the electrolysis model, teachers support students by providing the following shortened checklist during their investigation:

- I understand that developing more details in the model will help me make more valid conclusions about the system
- I have added mechanisms to the model
- I have had discussions with my group about any abstract variables and relationships that are not on my model
- I have revised the model based on my experiences with the electrolysis apparatus

To check for appropriate metacognition in this step, teachers should ask a few questions about the choices that students make when they perform the practice such as

- How did you identify abstract variables and relationships?
- Why is it important to consider all variables, relationships, and mechanisms in this electrolysis system?
- How do you know that the model you have developed will be valid?

If students struggle to answer these questions, then they likely have not mastered the learning tasks of the practice. Teachers can help students by going back and modeling the difficult practices, then giving students another chance to emulate. When students can answer these questions in a way that demonstrates their mastery of defining problems, then teachers can fade all support in the next step, self-reflection.

5.4.2 Self-Reflection

In the self-reflection step of MPI-S, students perform the targeted practice entirely on their own and reflect on the outcome. This last step of MPI-S is aligned to the self-reflection processes of SRL (Figure 5.1) and helps students to self-evaluate their performance and attribute their successes and failures to sources of learning. The self-reflection step builds upon the self-control step because students are expected to regulate their learning without any support. Students should be able to demonstrate that they can both understand and implement the science and engineering practice (developing and using models) without any teacher support. In this step, the teacher gives the learning task and ensures that the student was able to accomplish it in a way that illustrates the science and engineering practice.

Developing and Using Models

The teacher can decide if the student has mastered the science and engineering practice by evaluating student answers to the rationale. Questions that a teacher would ask regarding developing and using models for grades 9 through 12 for the development of the electrolysis model are:

- How important is the development of a model that has both observable and abstract features? Explain your response (Task value)
- Explain how you identified variables, relationships, and mechanisms of the electrolysis system (Metacognitive monitoring)
- What are some strengths and weaknesses of the model you developed? (Self-evaluation)
- How might you improve the weaknesses in the model? (Self-evaluation)
- What about developing and using models do you think you do well? Why? (Attribution)
- What about developing and using models do you think you still need to practice? Why? (Attribution)

In the next section, there will be two case studies to demonstrate how different processes of self-regulated learning could help or hinder students in their learning about developing and using models.

5.5 Teacher–Student Cases

5.5.1 Case Study: Adaptive

The case study featured in this chapter as an example of an adaptive and effective learner for developing and using models will focus on:

- Task value in the forethought phase
- Metacognitive monitoring in the performance phase
- Attributions in the self-reflection phase

Augusta, an 11th grade student, is working with her group to develop and refine a model of the variables, relationships, and mechanisms in an electrolysis system. She understands the goal she and her group are expected to achieve. Augusta is motivated to begin to think about the details of the system. She also realizes that there will be factors of the model that will be physically observable and that there are some that will be abstract, so she is ready to think about both kinds. Although she is not intending to go to college to study physical science, she embraces the challenge of trying to account for all variables, relationships, and mechanisms in an electrolysis system because she is open to learning new skills. She realizes that even though she cannot anticipate when she would use the skill of developing and using models in her adult life, she wants to leave public school with as much knowledge as possible. Her attitude is "You never know, these skills might come in handy someday" (task value).

Augusta took notes when her teacher was modeling how to develop the electrolysis conceptual model about the process her teacher used. She even checked her notes against the checklist her teacher gave her to monitor her progress. She jotted down a few of the process goals that were not on the checklist to help her when she started to add relationships and mechanisms in her conceptual model (emulation phase). When August was observing her group's electrolysis apparatus, she used her embellished checklist to be sure she was accounting for everything she could do to complete the model (metacognitive monitoring). She used the checklist several times during the observations phase and then double-checked her group's model against another group's model to see if they had all of the details they needed.

When August received her teacher's feedback on her work, she was pleased that it was mostly positive (self-reaction) and she noted the few things that she and her group did not note on the model. She felt that her success on developing the model was based on her ability to monitor her progress with her checklist (attribution). She also had a realization that monitoring her thinking with the checklists did not help her attain a perfect product, so she realized that she needed to do more than just monitor the checklist (attribution). She made a mental note that the next time they do a model, she was going to take a break from looking at the checklists to examine the model and see if she had any ideas outside of the checklists (adaptivity).

5.5.2 Case Study: Maladaptive

The case study featured in this chapter as an example of an ineffective learner that needs support for developing and using models will focus on:

- Task value in the forethought phase
- Metacognitive monitoring in the performance phase
- Attributions in the self-reflection phase

Hank, an 11th grader in Augusta's class, paid attention to the teacher when they modeled the development of the electrolysis conceptual model, but he did not record what he was expected to do. He figured that he was going to be a chef when he got out of school, so creating models was not important to him (lack of task value). Hank believed that even if he did not do well with developing and using models, it would not really affect his career plans, so he decided not to put a lot of effort into trying.

Hank went through the checklists in a very mechanical way. He was only partially aware of the outcome goal and the process goals, and he did not really take responsibility for learning them if they did not come easily to him. As a result, he followed the "recipe" for developing and using models from the checklist, but did not realize why he was doing the actions on the checklist.

Hank received feedback for his model and although it had some strengths, it had more weaknesses than his teacher would have liked to see. Hank received his feedback and thought, "wow, my teacher just does not like me, so I got a poor grade (self-reaction and self-evaluation)." In Hank's mind, this took care of any failures he had so he could move on without changing his processes (attribution). In this case, Hank attributed his failures to an (imaginary) external, uncontrollable source, so any adaptation to be more successful at developing and using models was unnecessary. This lack of adaptivity is likely going to affect Hank's next attempt at learning something similar and he probably has even lower task value for modeling than he did before the electrolysis activity.

5.5.3 Supports the Teacher Can Provide for Hank in This Case Study

5.5.3.1 Task Value
Hank did not realize the intrinsic value in developing and using models because he thought the skill was irrelevant for his career plans of going to a culinary academy after public school. The teacher, knowing a bit of Hank's background, could have approached him and related the features of electrolysis to the features of cooking. The teacher could have introduced him to websites that show people tinkering with recipes to find out the variables, relationships, and mechanisms that improve the quality of food preparation. For example, there are many different recipes for the "perfect" scrambled egg, and the recipes advocate for different processes. Having a model for the ingredients, relationship between the ingredients and mechanisms behind the reactions that happen to the food may help a chef adjust their process to improve the output. Hank may realize the value for the task once the teacher points it out.

5.5.3.2 Metacognitive Monitoring
Since Hank was only interested in surface learning, helping him see the value in learning how to develop and use models may increase his

metacognitive monitoring with the checklists. If he had more value in the task, Hank would likely want to master the skill because he could see how it was applicable to his future interests. Another approach the teacher could take would be to put Hank in a group and create roles for the group to justify how they accomplished the tasks involved in developing and using models. Placing Hank in a group that has required interdependence may support Hank doing more than going through the motions.

5.5.3.3 Attribution

Hank attributed his failures in modeling to an uncontrollable and unchangeable external force. In addition, Hank did not even consider his attributions for his successes in the modeling activities. The teacher could have some influence over Hank by helping him see what might be causes of his successes. The teacher could show that paying attention to her modeling helped Hank understand the checklists so that he had an awareness of how to get to his outcome. Having Hank explain how he accomplished the strong parts of the model could both help him set his attributions to internal and controllable sources, but also could help him find value and self-efficacy in the task. Once the teacher is comfortable that Hank is examining some of his internal processes, she could ask about the weaknesses in Hank's model. By comparing Hank's approaches to the strong and weak parts of his model, the teacher may be able to help Hank gain more awareness of his self-reflection processes.

5.6 Designing Lessons That Use the Practice of Developing and Using Models

In order to plan for teacher modeling of the practice of developing and using models, teachers can review a lesson plan for ways they can model the key learning tasks in an investigation. Teachers can look for opportunities in their lesson to point out to students:

- Identification of variables in a phenomenon, system, or design
- Relationships in a phenomenon, system, or design
- Mechanisms in a phenomenon, system, or design
- Ways to identify strengths and weaknesses in a model

Teachers can then make checklists for students by listing the key features of doing the science and engineering practice, and then convert them into student-friendly language for a checklist bullet. For example, a key feature from developing and using models is to evaluate a model for strengths and weaknesses and revise the weaknesses of the model. The checklist bullet item for this key learning task can be "When I see a weakness in a model, I try to learn more about that area of the phenomenon, system or design so I can revise it." Teachers should make at least one bulleted item for each key feature of the practice. Making more than one item could potentially help students by understanding the practice from different perspectives.

5.7 Teacher Reflection on Implementation of Lesson Featuring Development and Use of Models

Being a reflective practitioner is important for supporting student self-regulated learning. Teachers should consider the following questions for the practice of asking questions and defining problems:

- What worked well in supporting students' development and use of models?
- What did not work well in supporting students' development and use of models?
- How will I change this for next time I teach students to develop and use models?

CHAPTER 6

Planning and Carrying Out Investigations

Chapters 4 through 11 in this book have a parallel structure, each dealing with a particular science and engineering practice using the same approaches but with unique examples. In each chapter, the practice is dissected into distinct and clear learning tasks. These tasks are then examined within the context of a self-regulated learning cycle. Teacher examples for instruction and assessment within the SRL cycle are emphasized. A coaching strategy for supporting student SRL is explained using the example of a miniature golf course design challenge that involves planning and carrying out investigations. The tasks are reexamined in two case studies – one positive and one negative – to demonstrate how the learning strategies for SRL can be used by students. The chapter ends with questions for teacher reflection.

6.1 Learning Tasks in the Practice

The Next Generation Science Standards (NGSS Lead States, 2013) explain that both scientists and engineers work collaboratively with their colleagues to plan and carry out investigations, and that these investigations are characterized by a systematic design. Scientists and engineers analyze their own and other investigations, examining the ability for the procedures to obtain valid and reliable evidence that is related to the research question. Whereas scientists plan and carry out investigations in order to better understand a phenomenon found in the natural world, engineers conduct investigations to identify the effectiveness, efficiencies, and durability of a design.

These overall definitions have been further broken down into grade bands by NGSS of K-2 (aged 4 to 7), 3–5 (aged 8 to 11), 6–8 (aged 12 to 14), and 9–12 (aged 15 to 18). Later in this chapter, objectives are explained for the science and engineering practice of planning and carrying out investigations for the grade bands. The chapter begins with the learning tasks for students in grades K through 2 and build on the new practices that can be learned in the successive grade band (NGSS@NSTA, 2014). From there, the chapter breaks down the learning tasks involved in each grade band for the practice that can be used as process goals, which represents what students should be able to do to master the practice. The chapter also provides examples of how a teacher can support the learning tasks in an explicit and reflective way. Educational research on epistemic knowledge in science has shown that students are less likely to learn an aspect of the nature of science when taught implicitly, but students are more likely to learn when taught explicitly by the teacher and when given intentional opportunities to reflect on how the aspect was demonstrated (Khishfe & Abd-El-Khalick, 2002; Peters, 2012; Peters & Kitsantas, 2010; Peters-Burton, 2015). That literature forms the basis for practical ways to enhance student learning for science and engineering practices.

Breaking the practice down is not only important for helping students who may not have had exposure to doing science and engineering, but it is also important to situate the learning task in a cycle of self-regulated learning. In order for self-regulated learning to be productive, the learning processes must begin with a well-defined learning task. As explained in Chapter 3, goal setting is key to being successful with self-regulated learning, and the goal needs to be discrete and composed of smaller, more proximal process goals that lead up to the outcome goal. If the outcome goal is not distinct, it is difficult for a learner to know if they have achieved the goal.

Outcome goals for planning and carrying out investigations have several facets. One is being able to design a systematic procedure using algorithmic thinking to create efficiencies. The identification of variables is key to being able to take measurements to provide evidence, as well as the ability to plan and carry out a fair test of the variables. Planning and carrying out investigations is related to asking questions and defining problems because the research question or problem to be solved drives the planning of the investigation. The practice of planning and carrying out investigations is also related to developing and using models because both practices require the ability to identify variables. However, recall that in order for students to master the outcome goals, they also need process goals that are smaller steps to reaching the outcome goals. The following section articulates the process goals for planning and carrying out investigations by identifying key characteristics of the practice.

6.1.1 Process Goals for Planning and Carrying Out Investigations

The Next Generation Science Standards note that planning and carrying out investigations to answer questions or test solutions to problems in grades K through 2 builds on prior experiences and progresses to simple investigations, based on fair tests, which provide data to support explanations or design solutions (NGSS@NSTA, 2014). Student learning tasks for that standard are broken into the bullet points below:

- Explain how in prior experiences students have compared two situations and evaluate if it is a fair test
- Plan and carry out an investigation collaboratively
- Consider different ways to make observations and measurements to answer a research question
- Make observations or measurements to collect data in an organized way that can be used to make comparisons
- Plan and carry out an investigation that changes only one variable at a time to see the change in outcome

6.1.1.1 Explicitly and Reflectively Supporting Process Goals for Planning and Carrying Out Investigations

Educational research has demonstrated that students can more effectively learn when teachers explicitly and reflectively support learning tasks (Peters-Burton, 2015). It should be noted that explicitly teaching a practice does not mean lecturing about a practice. Rather, explicitly teaching a practice means to point out when the practice is being done so that students can notice the features of the behavior. Teachers can model the behaviors that are done by scientists and engineers when performing the practice of planning and carrying out investigations to help students see their outcome goal. When reflectively teaching the practice, teachers give students a chance to show what they know about performing the practice. From assessing the student's reflective response, teachers can determine if they need to reteach the practice or give students another chance to emulate the practice.

Teachers can help students in grades K-2 learn to *compare two situations and evaluate if it is a fair test* by modeling the skill explicitly. As a whole class, ask students to brainstorm how they have compared two situations from their prior experiences. Keep a list of the experiences and sort them into comparisons that were scientific (same conditions) and those that are not scientific. Supporting students in their own reflection on how well they performed the target skill can help them assess their proficiency and make changes in the future if needed. In small groups, have students take an example from the scientific (same conditions) list generated in the brainstorming activity. Have students discuss how to test this example in a fair way.

Teachers can be explicit about *planning and carrying out an investigation collaboratively.* Choose several students to be in a demonstration group with the teacher. Select a question to answer or problem to solve. Talk to students about what to do to be collaborative (active listening, take on roles, connect to other students' ideas when talking, taking turns). Make a T-chart to explain what collaboration in science looks like and sounds like. To help students be reflective about their collaboration skills, put students in small groups and give them copies of the T-chart that was made in the demonstration. Have students work collaboratively to plan and carry out another investigation. Have students make checkmarks after the action (looks like/sounds like) from the T-chart when they do the action in their group. Have a discussion with what was easy and what was difficult from the T-chart.

Teachers can help students to *consider different ways to make observations and measurements to answer a research question* by first being explicit about the behavior. In a whole class, ask students to think about how they would answer a question like "What makes an object sink or float in water?" Have them do a Think/Pair/Share to brainstorm ideas to make observations or measurements to answer that question. As students give their responses, talk aloud about how the observation might help to answer the research question. To help students be reflective about different ways to make observations and measurement, provide students a research question and ask them to list as many different ways to make observations or make measurements as they can to answer the research question.

Explicitly demonstrating how to *make observations or measurements to collect data in an organized way that can be used to make comparisons* can help students understand another facet of planning and carrying out investigations. As a whole class, discuss a research question and demonstrate or show a video of the investigation. Talk aloud about how you are making the observations or measurements and explain how you are recording the data in a way that is organized. Teachers can then help students reflect on their performance of being organized in making observations or measurements. Given a research question, ask students to construct a data table for collecting data. Check the table, give students feedback, and have them revise if needed. Ask students to collect data in the data table and have them reflect on the experience.

Teachers can demonstrate to students *how to plan and carry out an investigation that changes only one variable at a time to see the change in outcome* so that students can see the skill in action. Set up a demonstration investigation for the class that deliberately changes more than one variable and talk aloud about your thinking. Ask students if they can be sure about the results if they change more than one variable. Guide students to see that changing one variable at a time is the only way to see the relationships systemically. To help students be reflective about the skill, ask small groups of students to plan an investigation. Have students peer review other group's investigation designs with an emphasis on isolating one variable at a time.

6.1.1.2 Explicitly and Reflectively Supporting Process Goals for Planning and Carrying Out Investigations for Grades 3–5

According to the Next Generation Science Standards, planning and carrying out investigations to answer questions or test solutions to problems for students in grades 3–5 builds on their K-2 experiences and progresses to include investigations that control variables and provide evidence to support explanations or design solutions. A list of learning tasks that can serve as process goals for these science and engineering practices is as follows:

- Plan and carry out investigations that control for variables
- Plan and carry out investigations that consider the number of trials
- Evaluate methods in an investigation for fairness, controlling variables and number of trials
- Test two different models of the same proposed object, tool, or process to determine which better meets criteria for success

Teachers can explicitly teach students to *plan and carry out investigations that control for variables* so that students can see what this new skill involves. After reading a fictional book, have students select a problem to solve from the book. For example, building a house that does not blow down from the story "The Three Little Pigs." As a whole class, plan an investigation for one variable that would help a model house stay upright with a strong wind blowing, as the teacher talks aloud about their thinking to control for a variable. To help students be reflective about this process goal, ask students to identify other variables that would help a model house stay upright with a strong wind blowing. Ask them to plan an investigation with one of those variables. Ask them to carry it out and report on how they controlled for only one variable.

Teachers can help students see how to *plan and carry out investigations that consider the number of trials* by being explicit about their thinking while modeling this skill. Prepare data with three trials for each level of independent variable for two different investigations: (a) one that has a clear pattern such as how long it takes a ball to travel a distance given different angles on a ramp, and (b) one that has a pattern that is harder to see, such as how long it takes an ice cube to melt in different conditions. Show students the data, explain each research question and how the data were obtained. Then ask "Is three trials enough to have good evidence?" Guide students to see that something like the ice melting needs more trials because it is less of a clear pattern. Teachers can help students reflect on their skills and evaluate their mastery of considering the number of trials in an investigation. Each time students are expected to plan and carry out an investigation, have the student groups discuss how many trials would be needed for the evidence to be convincing.

To explicitly teach students to *evaluate methods in an investigation for fairness, controlling variables, and number of trials*, teachers can present a plan for an investigation to the class that has intentionally designed flaws. Have students in small groups discuss and evaluate their investigation for fairness, controlling variables, and number of trials. Have groups report out and guide students by discussing your thinking about these factors. After the class discussion on fairness, controlling variables, and number of trials, have students reflect and revise the investigation to improve all three of these factors to be reflective about their skills.

Teachers can explicitly demonstrate how *to test two different models of the same proposed object, tool, or process* to determine which better meets criteria for success to benefit the students. Present a design challenge to the class. Conduct a discussion on how to plan for a test of two different models of the same tool to determine which better meets criteria for success. To support student reflection on their skills, have student groups carry out the plan for a test of two different models of the same tool to determine which better meets criteria for success. Have students discuss their reflection on how the results helped them decide which condition met the criteria more fully.

6.1.1.3 Explicitly and Reflectively Supporting Process Goals for Planning and Carrying Out Investigations for Grades 6–8
According to the Next Generation Science Standards, planning and carrying out investigations for students in grades 6–8 builds on K-5 experiences and progresses to include investigations that use multiple variables and provide evidence to support explanations or solutions. The following

breakdown of learning tasks can be used as process goals for students learning this science and engineering practice:

- Identify independent and dependent variables when planning and carrying out an investigation
- Consider what tools are needed to carry out an investigation
- Decide what measurements need to be recorded and how to record data in a clear and consistent way
- Peer review and revise a plan for an investigation to improve methods
- Collect data to determine the performance of a tool, system, or process in a range of conditions

Teachers can help students identify independent and dependent variables when planning and carrying out an investigation by being explicit in their instruction about the skill. Present to the class a research question, the model for the phenomenon, and the investigation plan. Have students identify the independent and dependent variables and guide their thinking by talking aloud about how you make decisions about variables. In the next investigation that includes a model, have students identify the independent and dependent variables on the model before they plan for and carry out the investigation so that they can reflect on their progress in mastering the skill.

Explicitly teaching what it means to *consider what tools are needed to carry out an investigation* is a first step in helping students understand how scientists and engineers use the practices. Present a research question to the class. Show students at least three different tools or methods that can be used to measure the phenomenon in the investigation and describe the pros and cons of each. Put students into small groups to plan the investigation with a focus on what tool they would choose. Bring students back for a class discussion and guide them regarding tool selection by talking aloud about your choices and considerations. To help students reflect on their achievement in this skill, present a research question to the class or have students develop the research question and model. Have students research different tools or methods that can be used to measure the phenomenon in the investigation and describe the pros and cons of each. Put students into small groups to plan the investigation with a paragraph on why they chose their measurement tool for the investigation.

To perform this practice, students will also need to *decide what measurements need to be recorded and how to record data in a clear and consistent way.* Teachers can explicitly teach the skill by walking through the development of a data table, asking students to identify variables, then showing them how you would design a data table for that purpose. Explain any derived variables that you may need to put into the data table as well. To help students reflect on their ability to perform the skill, have students develop a data table after they plan for an investigation. Have students carry out the investigation, taking notes about how well their data table worked for organization and completeness. Have students write a reflection on what they learned about creating a data table.

To explicitly teach students to *peer review and revise a plan for an investigation to improve methods*, create a rubric for peer reviewing a plan for an investigation and discuss it with the class. Have students use the rubric to peer review the investigation and guide them in their thinking. Students can reflect on their performance in an active way as well for this skill. Based on what was decided in the peer review, have students revise the investigation and explain their reasoning.

Teachers can explicitly teach students to *collect data to determine the performance of a tool, system, or process in a range of conditions*. Demonstrate to students the pros and cons of collecting data with an analog and a digital tool of the same measure, such as two versions of a thermometer. Conduct a demonstration with the data being shared with students in real-time. Ask students to comment on a graphic organizer as data is being collected on the performance of the tool and after the data is collected. Have students report out what they recorded on their graphic organizer and guide them by talking aloud about your thinking on the evaluation. To help students be reflective about their skill, provide students a space to record their reactions to their data collection as a way to prompt them to be mindful about the performance of the tool, system, or process as they collect data.

6.1.1.4 *Explicitly and Reflectively Supporting Process Goals for Planning and Carrying Out Investigations for Grades 9–12*

Students in grades 9–12 should be able to perform key learning tasks about planning and carrying out investigations by building on K-8 experiences and progress their learning to include investigations that provide evidence for and test conceptual, mathematical, physical, and empirical models. The following learning tasks can serve as process goals for students in grades 9–12 for the science and engineering practice of planning and carrying out investigations:

- Carry out investigations and revise models based on the patterns observed in the data
- Carry out investigations and consider confounding variables. Revise plans for investigation to minimize confounding variables

- Consider the limitations for validity and reliability of data collected in a plan of an investigation. Revise investigation accordingly
- Consider environmental, social, and personal impacts when planning an investigation and revise plan accordingly
- Manipulate variables and collect data about a complex model of a proposed process or system to identify failure points or improve performance relative to criteria for success or other variables

Teachers can explicitly teach students how to *carry out investigations and revise models based on the patterns observed in the data* by presenting students with research question, model, investigation plan, and data table. Point out patterns that you see in the data and what made you notice them. Record the patterns and demonstrate to students how you would go back and alter the model based on the patterns you noticed in the data. Students can be reflective about their skills to evaluate their proficiency with the teacher's support. When students have developed a research question, model, investigation plan, and have carried out the investigation, have students discuss in their lab group what patterns they see in the data. After they reach consensus in their group, have students make changes on the conceptual model they constructed previously in a different color and explain how the patterns in the data led to these changes.

Teachers can explicitly teach students how *to carry out investigations and consider confounding variables and revise plans for investigation to minimize confounding variables* by providing an investigation for the class. Walk the students through the identification of confounding variables, how you identified them, and how the variables may influence the results in the current design. Map out how each of the confounding variables can be dealt with through design changes. When designing an investigation, teachers can help students be reflective by providing space for students to consider the confounding variables that they designed out of the investigation and for the possible confounding variables that they may find once they carry out the investigation. Have students discuss their notes on confounding variables as a class.

Teachers can explicitly teach students how to *consider the limitations for validity and reliability of data collected in a plan of an investigation and revise the investigation accordingly* to help students understand the skill. Given a particular investigation plan, demonstrate to students the possible points to shore up validity and reliability. Discuss aloud your decision points and what alerted you to the validity and reliability issues. Then help students be reflective about their skill. After students plan an investigation but

Planning and Carrying Out Investigations 97

before they carry it out, have other students peer review the plans, looking for limitations of validity and reliability. Have students offer suggestions for revisions before carrying out the investigation.

To help students consider *environmental, social, and personal impacts when planning an investigation and revise plan accordingly*, teachers can model the behavior explicitly. Given a particular investigation plan, ask students to form groups representing environmental impacts, another group representing social impacts, and another group representing personal impacts. Meet with each group and conduct their discussion and guide their thinking by discussing how you would consider the impacts. Bring all of the groups together to discuss their viewpoints and walk them through how to revise the investigation plan accordingly. Then help students to be reflective. After students plan for an investigation but before they carry it out, have them explain how they took into account environmental, social, and personal impacts. If they did not or if they missed any of the target impacts, have students revise the plan accordingly.

Students can learn to *manipulate variables and collect data about a complex model of a proposed process or system to identify failure points or improve performance relative to criteria for success or other variables* by observing a teacher model these behaviors. Have students bring in ideas for a design challenge for a process or system. Select one and collaboratively with the class, identify variables, and develop a model. Have the students carry out the design challenge and present their findings to the class for peer review with a focus on failure points and improvements. Ask students to explain their thinking and add details about the process when necessary. When students have defined a problem, created a complex model of the process or system, and planned and carried out an investigation of the process or system, have them be reflective by discussing their data patterns and identify failure points. Have them present improvements to the design for the next attempt at the design cycle.

6.2 Strategies for Teachers to Support Student Practices

For the practice of planning and carrying out investigations, this section will focus on the SRL processes of *goal orientation* in the Forethought phase, *attention focusing* in the Performance phase, and *adaptation* in the Self-Reflection phase as seen in Figure 6.1. The focus will be on these processes because they are most illustrative for the practices of planning and carrying out investigations. In other chapters discussing other practices, the book will focus on other SRL processes so that the whole cycle will be discussed across the book.

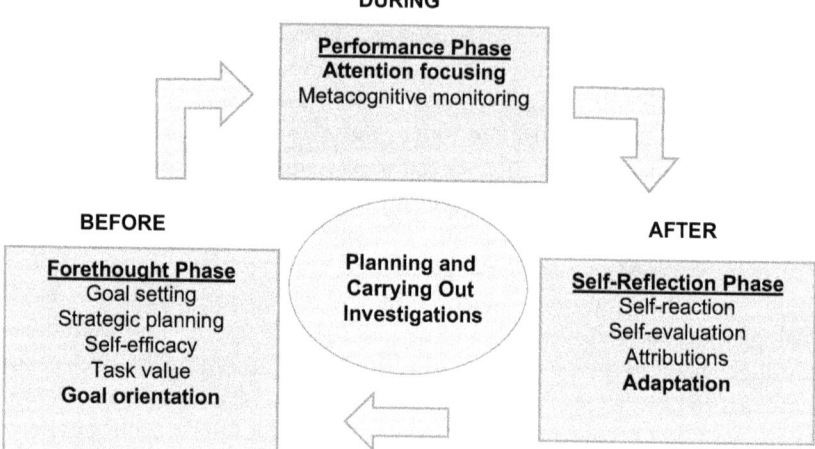

Figure 6.1 SRL processes for planning and carrying out investigations

6.3 Instructing Students in Self-Regulating Their Learning about Planning and Carrying Out Investigations

A research-based coaching strategy has been developed for teachers to demonstrate strategic SRL thinking, called Metacognitive Promoting Intervention-Science (MPI-S; Peters, 2009; Peters & Kitsantas, 2010). MPI-S enacts SRL because it prompts students about setting advantageous goals for learning a task by explicitly demonstrating the practice, which gives students who do not have experience with the practice an entry point to learning. MPI-S also assists students in focusing their attention on the key features of planning and carrying out investigations, and asks students to reflect on their successes and failures in their performance of the practice and adapt accordingly. This teaching strategy works well for lessons involving science and engineering practices because it engages students with the processes and approaches to thinking rationally and systematically.

MPI-S is a suite of curricular tools, made up of a suite of checklists and questions that can be incorporated into established lesson plans to support student SRL strategies. The implementation of MPI-S consists of four steps: Modeling, Emulation, Self-Control, and Self-Regulation. The steps of MPI-S are the same ones as the coaching strategy founded by Zimmerman (2000). In the modeling phase, the teacher demonstrates key features of planning and carrying out investigations. The student considers their own forethought processes. In the emulation phase, the teacher provides a checklist of key features of planning and carrying out investigations. Students use the checklist as a tool for forethought processes. In the self-control phase, the teacher provides a short checklist and asks questions about student learning strategies. The students monitor their progress for planning and carrying out investigations. In the self-regulation phase, the teacher asks students to explain how and why they used key features of planning and carrying out investigations. Students identify the instances of learning and assess their quality of planning and carrying out investigations.

Using this approach, the teacher initially supports students explicitly through modeling and then drops the level of support so that students are able to articulate how they understand how to plan and carry out investigations (or any of the science and engineering practices) independently. The first two steps of MPI-S (modeling and emulation) are instructional and the second two steps (self-control and self-regulation) assess student learning to inform instruction. MPI-S does not increase the time it takes to teach science and engineering practices, because as the teaching focuses on student use of science and engineering practices that are tangible at the same time, they are creating opportunities for students to be more aware of their learning strategies.

6.3.1 Modeling

Modeling is the first step in the MPI-S teaching strategy and is aligned to forethought processes of SRL, in this case goal orientation (Figure 6.1). Students who have a performance goal orientation may be averse to taking risks in their learning, which results in moving forward with the learning activity without appearing to fail or struggle, even if it is at the risk of not actually learning (Dweck & Leggett, 1988). Students who adopt a mastery approach to their goal orientation to learning are not afraid to fail or demonstrate publicly that they are struggling because their focus is to learn the task deeply.

The modeling step in MPI-S helps students with their forethought processes such as goal orientation because it demonstrates their learning goal, and helps them evaluate their self-efficacy and value in the task. In this case, students may see that when the teacher models when they struggle or find points of difficulty in learning yet persist by finding another learning approach, that it is not detrimental. Struggle in learning can produce even more advanced learning outcomes because the learner needs to persist and find other ways to achieve. Therefore, teachers should take care to explain other applications of the skill of planning and carrying out investigations so that students can have goal orientation beliefs about mastering the learning task. Teachers can also add other examples of valuable uses of the practice for students during later phases of the coaching strategy.

Modeling is much like a cognitive apprenticeship (Collins, Brown, & Newman, 1989) where the mentor (teacher) does the activities in full view of the apprentice (students), but at the same time talks aloud about rationale, choices, and decision points with the intention that the apprentice will be able to adopt the same practices. The role of the student in this first step of MPI-S is to notice key features of the skill as demonstrated by the mentor and ascertain the overall sense of the outcome. Students will learn how to reach the outcome through understanding process goals in later steps of the teaching strategy.

For example, the teacher may design an investigation in middle school (grades 6–8) that involves a design challenge for students to create one hole per group for a miniature golf course that consists of obstacles made up by each group that has at least one pendulum. A miniature golf course is a game played for entertainment by adults and children. It usually consists of nine to eighteen holes that have different features and turns to make getting the ball into the hole more difficult. The miniature golf course hole assignment requires at least one object swinging back and forth (a pendulum) as an obstacle to prevent the ball to reach the hole. Students are required to design the dimensions of the golf course hole and to give instructions for making the most optimal shot. From this context, students are expected to learn about the periodic motion of a pendulum.

Recall from the Next Generation Science Standards that some of the key features of planning and carrying out investigations that students in middle school are expected to master are:

- Identify independent and dependent variables when planning and carrying out an investigation
- Consider what tools are needed to carry out an investigation
- Decide what measurements need to be recorded and how to record data in a clear and consistent way
- Peer review and revise a plan for an investigation to improve methods
- Collect data to determine the performance of a tool, system, or process in a range of conditions

In the modeling step of the coaching model, teachers can demonstrate how they would go about designing a hole for a miniature golf course using principles of physics and geometry. The teacher could first give an overview of the design challenge by explaining the outcome goal.

In this design challenge, you will work in a group of four to design one hole for a miniature golf course. The course must have at least one turn that is more than 45 degrees and an obstacle that swings back and forth in front of the hole. The deliverable for the client will have three parts:

1. *Scaled plans for the hole*
2. *Prototype of the hole*
3. *Description of the optimal shot of the hole with investigation results from prototype*

The products will be peer reviewed for improvements before final submission to the client. The teacher can then model how they would begin the task. The teacher should anticipate to make a few mistakes in front of the class when they are modeling and explain how they recognize they made a mistake, and how they would change course to correct the mistake as best as they could. For example, when 1:10 sketching scaled plans for the hole, the teacher may notice that the overhead sketch for the final stretch of the hole looks too short to meet the criteria of the challenge. They can then go back and recheck the measurements of the scaled drawing in front of the class, talking aloud about what they see. The teacher could demonstrate to students that they miscalculated the scale to be 1:20 rather than 1:10 and change the scale. They could emphasize that they took the time to evaluate their drawing to see if it matched to their expectations. When the drawing did not match their expectations, they checked their calculations and realized it was a mistake and they took action to correct it before moving on.

In this way the teacher is modeling for constructive student goal orientation. The teacher showed how they cope with difficulties and failures when engaged in this learning task. Explicitly talking about what they learn from their mistakes, the teacher can foster the sense that a person is not a failure when they make a mistake, and that making a mistake is a learning opportunity. Students may be reluctant to show their mistakes to their peers and to the teacher, and modeling learning from mistakes can help students feel more comfortable to set a mastery goal orientation rather than being driven by appearing to be successful to an outside observer. Additionally, reducing a student's fear of failure will help them establish a mindset that is akin to an engineer's mindset. The engineering design process is focused on improvements, thus communicating that there are no expectations of perfection.

6.3.2 Emulation

It is during this second step of the coaching strategy when the shift from teacher-led to student-led activities begins. The emulation step is related to the SRL phase of forethought (Figure 6.1), but is different from modeling because it guides students to be the initiator of their learning about developing and using models. During emulation, the role of the student is to replicate the science and engineering practices that the teacher models when they are given a similar task as the model. However, the student does this with considerable support, and the teacher provides the students with a checklist to set goals, consider their self-beliefs regarding planning, and carrying out an investigation. The checklist should begin with the outcome goal for the investigation and then list the process goals that will help students define the problem they are going to pursue in the investigation.

In the emulation phase of the design challenge, students in small groups can brainstorm how their miniature golf course hole will look with a mind toward the criteria of having one turn greater than 45 degrees and having one obstacle that swings back and forth. Students are supported with a checklist of the following statements that have been addressed in the modeling step of the coaching strategy:

- I understand the expectations of the three deliverables required by the client in a general way (outcome goal)
- I understand that the group will need to decide on what to use as a pendulum for the obstacle
- I have identified the independent and dependent variables for tests of the pendulum obstacle
- The group has made a list of the tools needed to carry out the investigation
- The group has decided what measurements need to be recorded and how to record data in a clear and consistent way

Students use these statements to help them reinforce their goal setting and task value for developing and using models. Like cognitive apprenticeships, MPI-S helps students who may not have had prior access to ways of knowing in science and engineering by explicitly pointing out not just how to do the practice (procedural knowledge), but also why they are doing the practice (epistemic knowledge). In later lessons, teachers can use different checklists for different science and engineering practices.

6.4 Assessing Students in Self-Regulating Their Learning about Planning and Carrying Out Investigations

6.4.1 Self-Control

The self-control step of MPI-S is related to the performance phase of SRL (Figure 6.1) because it helps students monitor their performance and focus their attention on learning about science and engineering practices. Students engaged in MPI-S to this point have observed what they are supposed to be accomplishing through the teacher model (Modeling) and have attempted similar skills and knowledge with support from the teacher (Emulation). In the third step, the teacher continues to support student self-regulation of learning about science and engineering practices but reduces support to allow students to actively reflect on their metacognitive strategies. The third and fourth steps of this coaching approach can also be used to assess how well students are beginning to self-regulate their new skills. Teachers should provide students with a more difficult attempt at the skill they are trying to build and give students only a few basic standards from which to check.

In the miniature golf design challenge example, students will share the design, and prototype with the class. Additionally, students will share the investigation results for an optimal shot with the class. The class will then peer review and the group will revise a plan for an investigation to improve their design. Students are expected to take over more responsibility for learning and the teacher acts as the facilitator by proving basic support, only intervening when misconceptions arise. Using the example of the miniature golf hole design, teachers support students by providing the following shortened checklist during their improved investigation:

- I understand the suggestions that were made during the peer review
- I have adjusted the prototype based on the suggestions for another round of investigations on the improved design
- I have revised the plan for the investigation based on the peer review

To check for appropriate metacognition in this step, teachers should ask a few questions about the choices that students make when they perform the practice such as:

- How did you identify the independent and dependent variables for your investigation?

- How do you know your measurements are accurate?
- Do you think someone who is not involved in this investigation can understand your data collection? Why or why not?

If students struggle to answer these questions, then they likely have not mastered the learning tasks of the practice. Teachers can help students by going back and modeling the difficult practices, then giving students another chance to emulate. When students can answer these questions in a way that demonstrates their mastery of defining problems, then teachers can fade all support in the next step, self-reflection.

6.4.2 Self-Reflection

In the self-reflection step of MPI-S, students perform the targeted practice entirely on their own and reflect on the outcome. This last step of MPI-S is aligned to the self-reflection processes of SRL (Figure 6.1) and helps students to self-evaluate their performance, attributing their successes and failures to sources of learning. The self-reflection step builds upon the self-control step because students are expected to regulate their learning without any support. Students should be able to demonstrate they can both understand and implement the science and engineering practice (planning and carrying out investigations) without any teacher support. In this step, the teacher gives the learning task and ensures that the student was able to accomplish it in a way that parallels the science and engineering practice. The teacher can decide if the student has mastered the science and engineering practice by evaluating student answers to the rationale. Questions that a teacher would ask regarding planning and carrying out investigations for the middle school students engaged in the miniature golf hole design challenge are:

- In what ways did you learn from your mistakes and peer review suggestions? (Mastery goal orientation)
- Explain how you identified dependent and independent variables in your investigation. (Attention focusing)
- Explain how you identified measurement tools to use in your investigation. (Attention focusing)
- Explain how you developed and used a way to record your data in your investigation. (Attention focusing)
- After peer review, what did you change in your design? Did it improve your results for the investigation? (Adaptation)

- What about planning and carrying out investigations do you think you did well? Why? (Attribution)
- What about planning and carrying out investigations do you think you still need to practice? Why? (Attribution)

In the next section, there will be two case studies to demonstrate how different processes of self-regulated learning could help or hinder students in their learning while they plan and carry out investigations.

6.5 Teacher–Student Cases

6.5.1 Case Study: Adaptive

The case study featured in this chapter as an example of an adaptive and effective learner for defining problems will focus on:

- Goal orientation in the forethought phase
- Attention focusing in the performance phase
- Adaptation in the self-reflection phase

Pepper, an 8th grade student, is working with her group to design the pendulum part of the miniature golf hole project. Pepper has played miniature golf previously so she has a good idea of what the outcome should look like. However, she realizes that this is her first time working with pendulums and calculating the period of the pendulum in order to explain the optimal shot for the hole. Pepper understands that she will likely make mistakes but is fine with making mistakes publicly (mastery goal orientation), because in the past she has learned even more from making mistakes than she did when she was successful on the first try. She does want to get a good grade from the project, but also realizes that it will not likely go perfectly on the first try. She feels up for the challenge.

When her teacher modeled how to identify dependent and independent variables in the pendulum investigation for the golf course hole, Pepper noticed that the dependent variable was always the one that you were curious about, and decided to use a mnemonic device of "it depends" to remember that the dependent variable is the one that is unknown (attention focusing). She also noticed that the teacher thought through the investigation and jotted down the variables in a table to make sure that all variables were accounted for in the table (attention focusing). She mimicked the teacher's behavior for her own data table construction.

Pepper asked another group to peer review her data table. Instead of waiting for the group to give her summative remarks on the data table, Pepper asked the peer review group to talk out loud about their evaluation of her group's data table, so she could see how their thinking was different from her own (adaptive thinking). She also encouraged her whole group to observe the peer review group's evaluation discussion. The peer review group discussed the pros and cons of two organizational items: putting the independent variable column in the table before the dependent variable and putting the derived variables in the table immediately after the relevant measured variables. Pepper was thinking that the dependent variables were more important, so she put them first in the data table. After the peer review discussion, she saw that it made more sense to put the variables in chronological order. She also saw the value in putting the derived variable column directly after the relevant variable columns. She often made the mistake of taking the numbers from the wrong column to calculate the derived variable. If Pepper had not listened to the peer review discussion and only taken the final feedback of the revised table, she might not have understood why they were suggesting those changes and would have been less likely to edit her table for clarity.

6.5.2 Case Study: Maladaptive

The case study featured in this chapter as an example of an ineffective learner that needs support for asking questions will focus on:

- Goal orientation in the forethought phase
- Attention focusing in the performance phase
- Adaptation in the self-reflection phase

Stormi, an 8th grader in Pepper's class, was often shy and usually did not speak up when she worked in groups. For the miniature golf hole project, Stormi did not want anyone to think she was confused (performance goal orientation), so she just watched the other members of the group and she copied their behaviors. Stormi was invested in not appearing to struggle in front of the teacher and her peers, so she did not ask questions or challenge herself during the pendulum investigation. She was happy to settle into a record keeping role, and the other group members did not notice whether she understood the concepts or not.

Because Stormi was putting her efforts into trying not to make any mistakes publicly, she was going through the motions when identifying the variables in a pendulum, creating a table, and carrying out the investigation. She copied her group members and did not try to check if she really knew why they were making decisions (lack of attention focusing). She was appearing to follow along, but it was only a coping mechanism rather than mastery of the skills. She had a complete data table, just like the rest of her group, but if asked, she would not have been able to explain which variables were independent and dependent or have been able to recreate a data table for the investigation on her own.

As a result of mechanically designing and carrying out an investigation, Stormi did not consider the importance of reflecting on her performance. Rather, she was thrilled when she received a good grade because her group had done well in the miniature golf project. Stormi did not leave the learning task with any additional awareness of her learning processes because she did not know what to look for in the reflection. Because she received the outcome she was hoping for, she did not feel the need to change any of her approaches to learning (performance goal outcome).

6.5.3 Supports the Teacher Can Provide for Stormi in This Case Study

6.5.3.1 Goal Orientation

Helping students to learn how to master skills and knowledge (set mastery goals) has two components. First, the teacher should work to explicitly set up an environment where students feel safe to make mistakes. A teacher can do this by making mistakes during modeling, explaining how they realized it was a mistake and how they would adapt their work to correct it. If any students have negative reactions toward making mistakes, the teacher could talk to the class about how making mistakes is part of learning and that engineers and scientists make many mistakes, but the important thing is to learn from them. Teachers can also have displays around the classroom showing how successful scientists and engineers have made mistakes, to make the behavior part of the learning culture. Second, the teacher can observe students while they conduct investigations and have one-on-one conversations with those who demonstrate difficulty in making mistakes publicly. Some students may need additional coaching on feeling more confident about making mistakes and working through the mistake to learn what the difficulty was.

6.5.3.2 Attention Focusing

Ideally, Stormi would have noticed the key features of planning and carrying out investigations from when the teacher modeled the behaviors and from the checklists the teacher provided. However, because Stormi was focused on her outward appearance of success, she did not pick up the key features of the science and engineering practice from the coaching approach. In this case, the teacher may have been able to choose one of the easier skills and had a one-on-one discussion with Stormi about her approaches to learning the skill. In this one-on-one discussion, the teacher could point out to Stormi the ways that she focused on key characteristics. In addition, providing an environment where students feel safe to make mistakes could encourage students like Stormi to have discussions with their peers or the teacher when they are struggling, rather than keeping quiet because they do not want to look like they do not know the learning task.

6.5.3.3 *Adaptation*

Adaptation can take two different forms, help seeking or changing behavior. If a learner recognizes difficulties with their learning approach during self-reflection, then they can change their behavior if they know another approach that may be successful. If the learner does not have in mind a way to change to a more successful behavior, then they may seek help from another person or learning resource. Supporting Stormi to adapt her learning behaviors in this case may take multiple approaches and multiple trials, particularly because Stormi received a positive outcome from her current behaviors (a good group grade). Stormi's teacher may want to have a consultation with Stormi to see what learning resources she used from the modeling and emulation coaching phases. Once that is established, the teacher can have continued conversations with Stormi about what she is using from the coaching phases to be more mastery-oriented in her learning approaches. The teacher may also encourage Stormi to document the learning resources she uses from the coaching and demonstrate to Stormi that she is using more learning resources. Hopefully, that will help Stormi make the connection that learning is an internally driven activity and she needs to take more responsibility for learning to be truly successful.

Question for Teacher Reflection: Reflect on the adaptive and maladaptive case study you just read and answer the following question. What student supports for asking questions can you use in your classroom from these case studies?

6.6 Designing Lessons That Use the Practice of Planning and Carrying Out Investigations

In order to plan for teacher modeling of the practice of planning and carrying out investigations, teachers can review a lesson plan for ways they

Planning and Carrying Out Investigations 109

can model the key learning tasks for this practice in an investigation. Teachers can look for opportunities in their lesson to point out to students:

- Identification of dependent and independent variables
- Ways to design data tables
- Identification of confounding variables in an investigation
- How to determine the number of trials for levels of independent variables

Teachers can then make checklists for students by listing the key features of doing the science and engineering practice, and then convert them into student-friendly language for a checklist bullet. For example, a key feature from planning and carrying out investigations is to decide what measurements need to be recorded and how to record data in a clear and consistent way. The checklist bullet item for this key learning task can be "When I am designing a data table, I think through the procedure and note what measurements need to be taken, then I evaluate how to organize the columns and rows." Teachers should make at least one bulleted item for each key feature of the practice. Making more than one item could potentially help students by understanding the practice from different perspectives.

6.7 Teacher Reflection on Implementation of Lesson Featuring Planning and Carrying Out Investigations

Being a reflective practitioner is important for supporting student self-regulated learning. Teachers should consider the following questions for the practice of planning and carrying out investigations:

- What worked well in supporting students planning and carrying out investigations?
- What did not work well in supporting students planning and carrying out investigations?
- How will I change this for next time when I teach students to plan and carry out investigations?

CHAPTER 7

Analyzing and Interpreting Data

As in the other chapters about science and engineering practices (Chapters 4 through 11), this chapter focuses on the practices of analyzing and interpreting data. In each chapter, the practice is dissected into distinct and clear learning tasks which can be used as process goals. These tasks are then examined within the context of a self-regulated learning cycle and points for instruction and assessment are emphasized. The tasks are reassembled into two case studies – one positive and one negative – to demonstrate how the learning tasks can be used by students.

7.1 Learning Tasks in the Practice

The Next Generation Science Standards (NGSS Lead States, 2013) explain that both scientists and engineers analyze and interpret data. Scientists analyze and interpret data related to natural phenomena in order to make meaning of the parts, interactions, and mechanisms. Engineers analyze and interpret data to explain outcomes of design tests. Both science and engineering practices use analytical and interpretive tools such as visualizations, statistics, and tabulation to uncover patterns that may not be initially observed in the data. Scientists and engineers are mindful of error and degree of certainty while analyzing and interpreting data.

These overall definitions have been further broken down into grade bands by NGSS of K–2 (aged 4 to 7), 3–5 (aged 8 to 11), 6–8 (aged 12 to 14), and 9–12 (aged 15 to 18). The following sections in this chapter explain each new objective for the science and engineering practice of analyzing and interpreting data in the context of the grade bands. The chapter begins with the learning tasks from grades K–2 and build on the new practices that can be learned in the successive grade band (NGSS@NSTA, 2014). From there, the chapter breaks down the learning tasks involved in each objective for the practice, which represents what students should be able to do in order to master the practice.

Analyzing and Interpreting Data 111

The following sections also provide examples of how a teacher can support the learning tasks (process goals) in an explicit way and in a reflective way. Educational research on epistemic knowledge in science has shown that students are less likely to learn an aspect of the nature of science when taught implicitly, but students are more likely to learn epistemically when taught explicitly by the teacher. Giving students intentional opportunities to reflect on how well they demonstrated the learning was also effective (Khishfe & Abd-El-Khalick, 2002; Peters, 2012; Peters & Kitsantas, 2010; Peters-Burton, 2015). This literature offers practical ways to enhance student learning for analyzing and interpreting data.

Breaking the practice down is not only important for helping students who may not have had exposure to doing science and engineering, but it is also important to situate the learning task in a cycle of self-regulated learning. In order for self-regulated learning to be productive, the learning task must be well-defined. As explained in Chapter 3, goal setting is key to being successful with self-regulated learning, and a productive goal should be discrete and composed of smaller, more proximal process goals that lead up to the outcome goal.

Key characteristics that can be considered outcome goals for analyzing and interpreting data include being organized and systematic, seeking patterns and trends in the data, and making meaning from the data relative to the research question. However, recall that in order for students new to these practices to master them, students also need process goals that are smaller steps to reaching the outcome goals.

7.1.1 *Process Goals for Analyzing and Interpreting Data*

Educational research has demonstrated that students can more effectively learn when teachers explicitly and reflectively support learning tasks. Explicit instruction does not mean lecturing the students. Rather, teachers can teach explicitly by pointing out when a student is thinking or acting like scientists and engineers. Teachers can model the behaviors that are demonstrated by scientists and engineers when performing the practice of analyzing and interpreting data in order to explicitly teach it (Peters-Burton, 2017). When reflectively teaching the practice, teachers give students a chance to show what they know about performing the practice. After assessing the student's reflective response, teachers can determine if they need to reteach the practice or give students another chance to emulate the practice.

7.1.1.1 *Explicitly and Reflectively Supporting Process Goals for Analyzing and Interpreting Data for Grades K–2*

In grades K–2, the Next Generation Science Standards have the expectation that students analyzing data build on prior experiences and progress to collecting, recording, and sharing observations (NGSS@NSTA, 2014). This standard can be broken down into the following process goals:

- Record information in an organized way that can be understood by other people
- Explain observations, drawn or written, to other people
- Use observations to describe patterns that answer scientific questions or solve engineering problems
- Compare predictions to observations

Teachers can explicitly model how to *record information in an organized way* that can be understood by other people by selecting an everyday phenomenon and asking students to make a prediction. For example, will a toy car travel faster, slower, or at the same speed when on a ramp that has a higher tilt? Prepare a chart with the three choices and have students make a mark or put a sticker on the chart to represent their prediction. Students should place their marks in a straight vertical line starting at the bottom, which forms a bar chart of their predictions. The teacher can then demonstrate to students how the chart can be read to show the results of the predictions. Teachers can support students reflectively by having students perform the investigation to answer the question, "Will a toy car travel faster, slower, or the same speed when on a ramp that has a higher tilt?" Guide them to design the same type of chart to record trials of the time it takes the car to go down a ramp at three levels. Have students record their information from the investigation.

Teachers can support student development of *explaining observations, drawn or written, to other people* by providing students with a drawing or written observations. Ask them what they notice on the observations and guide them to make systematic observations by talking aloud about what a scientist or engineer would notice on the observations. Teachers can help students be reflective about their skill by providing the latter with a new drawing or written observations. Ask students to make observations individually and pair up to explain what observations they made. Conduct a class discussion about what observations can be made.

Teachers can help students *use observations to describe patterns that answer scientific questions or solve engineering problems* explicitly by providing students with a drawing or written observations and a question to answer or a problem to solve. The teacher can demonstrate what they are looking at on the observations and how they are recognizing patterns that are connected with the question or problem. Teachers can help students be reflective by providing students with a new drawing or written observations with a question or problem to solve. Ask them to make observations and describe patterns in the observations individually. Students should then pair up to explain what observations they made and the patterns that they saw in the observations. Conduct a class discussion about what observations can be made and patterns that were seen in the observations.

Teachers can help students learn the practice by modeling how to *compare predictions to observations*. Teachers can present a prediction and an observation to the class. Accompanied by a class discussion, make a Venn diagram showing how predictions are different and how they are the same as observations. Then teachers can help students be reflective by providing students with copies of the Venn diagram comparing predictions and observations. Provide students with a list of predictions and observations and ask them to use the Venn diagram to distinguish them.

7.1.1.2 Explicitly and Reflectively Supporting Process Goals for Analyzing and Interpreting Data for Grades 3–5
According to the Next Generation Science Standards, students analyzing data in 3–5 build on experiences from grades K through 2 and progress to introducing quantitative approaches to collecting data and conducting multiple trials of qualitative observations. When possible and feasible, digital tools should be used. The following standard can be broken down into process goals that break down the standard into learning tasks:

- Represent data in tables and explain patterns
- Represent tables in graphs and explain patterns
- Use logical reasoning, mathematics, and computation to make sense of phenomena
- Compare data collected by different groups and discuss similarities and differences
- Use data to refine design solutions

Teachers can help students learn to *represent data in tables and explain patterns* through explicit instruction. Conduct an investigation as a demonstration, focusing on gathering data and explaining the patterns you see while gathering data. While asking students questions about gathering data and explaining patterns, explain how you make decisions to place data in the table and how you locate patterns. Teachers can then help students be reflective by providing students with a data table for an investigation. Have students work in small groups to place data in the table correctly while the teacher walks around checking. When data is collected ask student groups to describe the pattern(s) they see in the data and why they made that decision.

Teachers can explicitly model the *representation of tables and graphs and explain patterns* by providing students with data that has already been collected in a table. While asking students questions about the data to be sure they understand the table, explain how you make decisions to select a type of graph as you project the graph to the students. Explain how you interpret the graph. Teachers can support student reflection by providing students with data in a table from an investigation. Have students work in small groups to select the correct type of graph and to construct the graph correctly while the teacher walks around checking. When the graph is constructed ask student groups to describe the pattern(s) they see on the graph and why they made that decision.

Teachers can explicitly model using *logical reasoning, mathematics, and computation to make sense of phenomena* by presenting a model for the phenomenon of the investigation that was demonstrated but without the new information from the data. Explain to students how you would add the new information from the investigation to the model while talking aloud about missteps and corrections along the way. Teachers can support student reflection by asking students to discuss the logical reasoning that you used in the model in small groups. Have a whole group discussion about the logical reasoning that was used in the model for the phenomenon.

Teachers can explicitly show students how to *compare data collected by different groups and discuss similarities and differences* to help reinforce this practice. From actual student data or from derived data, provide students with two sets of data collected for the same investigation that is somewhat different. Demonstrate to students what you see in the data sets and what questions you would ask the people who did the investigations to find the source of the differences. When an investigation is complete, teachers can help students be reflective by having students present their data and/or graph of the results and explain what they see. Have the other groups compare what they have for results and have them discuss any reasons for differences in the results. Typically, the differences come from slight alterations in the procedures.

Explicitly teaching students to *use data to refine design solutions* can begin by providing students with a prototype and data that investigates the effectiveness of the design solution. Ask students what they would change and why, while guiding their statements with your own thinking. To set up students for reflection of the use of data to refine design solutions, in an engineering design cycle, have students collect data on their design solution and present to the class for peer review. After peer review, have the group explain how they would refine their design solution based on the data and on the peer review suggestions.

7.1.1.3 Explicitly and Reflectively Supporting Process Goals for Analyzing and Interpreting Data for Grades 6–8

The practice of analyzing and interpreting data in grades 6–8 builds on student K–5 experiences and progresses to extending quantitative analysis to investigations, distinguishing between correlation and causation, and basic statistical techniques of data and error analysis. This practice can be broken down into process goals to help students better understand what it means to perform analyzing and interpreting data:

- Identify linear and nonlinear relationships in large data sets and/or graphical representations of data
- Identify temporal and spatial relationships in large data sets and/or graphical representations of data
- Distinguish between causal and correlational relationships in data
- Connect variables, relationships, and mechanisms in a phenomenon or design solution to data trends
- Apply appropriate statistical techniques to find central tendency and variability in data set, digitally when possible
- Consider errors and limitations in data set
- Consider ways to improve precision and accuracy with automation or improved methods

Teachers can model the *identification of linear and nonlinear relationships in large data sets and/or graphical representations of data* explicitly by providing students with a large data set and asking them questions about how they might find patterns in the data. Guide their ideas by thinking aloud about how you would go about identifying relationships (linear or nonlinear) by using a graph. Teachers can then help students be reflective about mastery of the practice by providing students with a large data set and the related question or problem to solve. Ask them to identify the relationship(s) found in the data. Students can discuss their process with the class or with the teacher.

Teacher modeling can assist students in *identifying temporal and spatial relationships in large data sets and/or graphical representations of data.*

Provide students with a large data set and ask them questions about how they can identify relationships involving either time or space in the data. Guide their ideas by thinking aloud about how you would go about identifying temporal or spatial relationships, making planned missteps, and correcting your mistakes aloud. Teachers can set the stage for student reflection of the practice by providing students with a large data set and the related question or problem to solve. Ask them to identify the temporal or spatial relationship(s) found in the data. Students can discuss their process with the class or with the teacher.

Teachers can assist students in learning how to *distinguish between causal and correlational relationships in data* by explicitly modeling the skill. As students are progressing through investigations across the year, make a note to reflect on the results and discuss if the type of methods for the investigations warrant a causal or correlation relationship. After students understand how to perform the practice, teachers can help students reflect on their work. On each investigation a student does during the year, set up questions in the conclusion portion of the investigation for the student to explain if the type of methods for the investigations warrant a causal or correlation relationship or not.

Explicit modeling of how to *connect variables, relationships, and mechanisms in a phenomenon or design solution to data trends* will demonstrate to students the goals they need to achieve to master the skill. Select investigations during the year to work as a whole class on the data analysis. Typically, these investigations may be introducing new analytical techniques. When working on the analyses as a group, begin with identifying variables and patterns from the results to describe relationships between the variables. Talk aloud about how you are seeing patterns in the data and interpretation of the relationships. If a mechanism is identifiable in the results, then explain how you are connecting the results to infer the mechanism. Once students feel they understand and have attempted the skill, teachers can help students reflect on their performance. In the conclusion section of an investigation report, ask students to explain how they identified relationships across the variables by noting the data they analyzed to infer the relationship. Similarly, ask students to identify the mechanism of the phenomenon and how they inferred that from the data.

Applying appropriate statistical techniques to find central tendency and variability in data set, digitally when possible can be modeled by the teacher so that students understand how to do the skill. After students have collected data for an investigation, conduct a workshop for them on the meaning of mean, median, and mode and how to calculate it digitally, if possible. Show students the results of all of the central tendencies of the data and explain which one is most appropriate given the research question, talking aloud about your reasoning. Explain the different variance techniques (digitally, if possible) in a similar way. In the analysis section of an investigation, ask students to be reflective by explaining which central tendency technique (mean, median, or mode) is most appropriate and why. Similarly, ask students to explain how they calculated variance and what it means with this particular data set.

Teachers can be explicit about how to *consider errors and limitations in data set* so that students perform the skill similar to scientists and engineers. After students have collected data for an investigation, conduct a workshop (a short, focused direct instruction) for them on the meaning of errors in investigations. Model for them how you would think about errors and limitations for this investigation by talking aloud about your decisions when analyzing the data for errors. In an investigation analysis section, ask students to reflect on the proficiency of their skill by explaining the limitations of the investigation after discussing the possible limitations with their group.

Teachers can explicitly teach how to *consider ways to improve precision and accuracy with automation or improved methods* by having half of the students in your class collect data for a particular research question and investigation in a traditional way, such as taking temperature with an alcohol thermometer. Have the other half of the class collect data with an automated tool, such as an electronic thermometer. Display the data sets to the students and calculate central tendency and variance while talking aloud about your process. Have a discussion with students about the precision and accuracy of the two different methods. Teachers can help students reflect on this skill as well. When assigning an investigation, have students select their tools to measure (a prior science and engineering practice learning task). Ask them why they selected that type of tool and the implications, both positive and negative, of selecting that type of tool on the precision and accuracy of the data. For example, an electronic thermometer may be more accurate, but may also display digits to the 10,000th place that could be confusing when calculating derived variables.

7.1.1.4 Explicitly and Reflectively Supporting Process Goals for Analyzing and Interpreting Data for Grades 9–12
According to the Next Generation Science Standards, students in grades 9–12 build on their K–8 experiences in analyzing and interpreting data

and progress to introducing more detailed statistical analysis, the comparison of data sets for consistency, and the use of models to generate and analyze data. The following learning tasks can be used as process goals to help students understand the details of this practice:

- Compare data sets from different groups and peer review the anomalies found between data sets
- Apply inferential statistics to answer questions or explain design solutions
- Evaluate the impact of new data on a working explanation or model

Teachers can explicitly teach students to *compare data sets from different groups and peer review the anomalies found between data sets* to help them see how the practice is performed in a systematic way. As a whole class, show students two sets of data from an investigation, one that has anomalies and one that does not. Ask students what they notice about the data sets while guiding them by modeling what you notice about anomalies and what questions you may have about the data collection procedures. To help students be reflective about the skill, at the end of an investigation, have students report on their data and analysis to the class for a peer review. When one group has different data from others, allow students to have a whole class discussion about what might have caused the differences and what should be done to either correct or amend the data. The difference may be due to errors or due to a different interpretation of the methods of data collection.

To explicitly support student *application of inferential statistics to answer questions or explain design solutions*, teachers can model the behavior. When introducing an investigation that has a new statistical technique, explain to students that you will be doing a workshop when they are finished collecting data. When they have their data sets complete, use a teacher-generated data set to apply inferential statistical techniques in front of the class while talking aloud to model your thinking and choices. Then show how this technique can answer the question or explain the design solution while talking aloud about what ideas you are noticing and connecting to the results. Then to be reflective, in a relevant investigation, students should apply inferential statistics to answer questions or explain design solutions and explain why they made the choices they did for the statistical techniques.

Teachers can also model the *evaluation of the impact of new data on a working explanation or model* to demonstrate to students how scientists and engineers go about the practice. At the end of a chosen investigation, have a workshop with the whole class on how to revisit their conceptual model (prior science and engineering practice) and incorporate the findings from the new data on the model. Talk aloud about what you are noticing and how you are making decisions to edit the model. Then to help students to be reflective, at the end of the analysis portion of an investigation, ask students to revisit their conceptual model that they started with in the investigation and make changes in a different color. Have student groups pair up to explain to each other the changes they made in the model.

7.2 Strategies for Teachers to Support Student Practices

For the practice of "analyzing and interpreting data," this section will focus on the SRL processes of *goal setting* and *strategic planning* in the Forethought phase, *attention focusing* in the Performance phase, and *adaptation* in the Self-Reflection phase as seen in Figure 7.1. The focus will be on these processes because they are most illustrative for the practices of analyzing and interpreting data. In other chapters discussing other practices, the book will focus on other SRL processes so that the whole cycle will be discussed across the book.

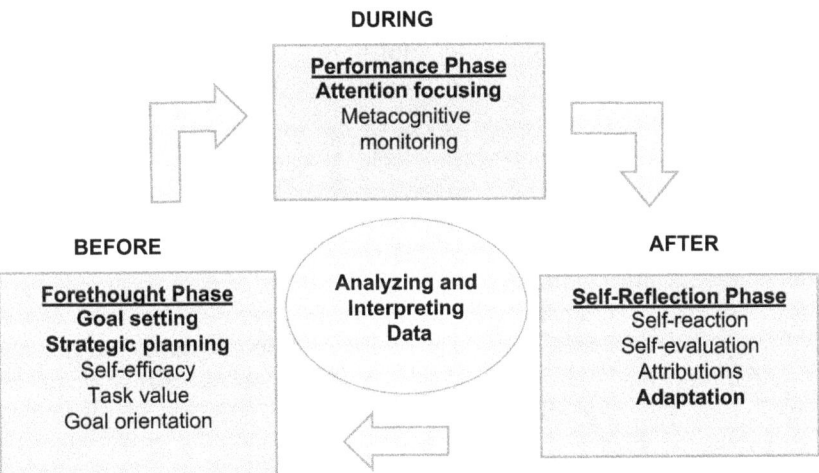

Figure 7.1   SRL processes for analyzing and interpreting data

7.3 Instructing Students in Self-Regulating Their Learning about Analyzing and Interpreting Data

Teachers can employ a coaching strategy that has been developed for them to demonstrate and support strategic SRL thinking, called Metacognitive Promoting Intervention-Science (MPI-S), which is based on a research-based teaching strategy (Peters, 2009; Peters & Kitsantas, 2010). MPI-S enacts SRL because it prompts students about setting advantageous goals for learning a task by explicitly demonstrating the practice, which gives students who do not have experience with the practice an entry point to learning. MPI-S also assists students in focusing their attention on the key features of analyzing and interpreting data, and asks students to reflect on their successes and failures in their performance of the practice and adapt accordingly. This teaching strategy works well for lessons involving science and engineering practices because it engages students with the processes and approaches to thinking rationally and systematically.

MPI-S is a suite of curricular tools, made up of a suite of checklists and questions that can be incorporated into established lesson plans to support student SRL strategies. The implementation of MPI-S consists of four steps: Modeling, Emulation, Self-Control, and Self-Regulation. The steps of MPI-S are the same ones as the coaching strategy founded by Zimmerman (2000). In the modeling phase, the teacher demonstrates key features of analyzing and interpreting data. Students consider their own forethought processes. In the emulation phase, the teacher provides a checklist of key features of analyzing and interpreting data. Students use the checklist as a tool for forethought processes. In the self-control phase, the teacher provides a short checklist and asks questions about student learning strategies. Students monitor their progress for analyzing and interpreting data. In the self-regulation phase, the teacher asks students to explain how and why they used key features of analyzing and interpreting data. Students identify the instances of learning and assess their quality of analyzing and interpreting data.

Using this approach, the teacher initially supports students explicitly through modeling and then drops the level of support so that students are able to articulate how they understand how to analyze and interpret data (or any of the science and engineering practices) independently. The first two steps of MPI-S (modeling and emulation) are instructional and the second two steps (self-control and self-regulation) assess student learning to inform instruction. MPI-S does not increase the time it takes to teach science and engineering practices, because MPI-S coaching focuses on student use of science and engineering practices and at the same time, creating opportunities for students to be more aware of their learning strategies.

7.3.1 Modeling

Modeling is the first step in the MPI-S teaching strategy and is aligned to forethought processes of SRL – in this case goal setting and strategic planning (Figure 7.1). Since students are often underexposed to the ways scientists think and conduct their work (Hogan, 2000), it is important that students begin a learning task by understanding the goal they are trying to reach. Often, students will be unable to set goals for themselves, so the teacher should step in and demonstrate what the student should be striving for. The modeling step in MPI-S helps students with their forethought processes because it demonstrates their learning goal, and helps them evaluate their self-efficacy and value in the task. Modeling is much like a cognitive apprenticeship (Collins, Brown, & Newman, 1989) where the mentor (teacher) does the activities in full view of the apprentice (students), but at the same time talks aloud about rationale, choices, and decision points with the intention that the apprentice will be able to adopt the same practices. The role of the student in this first step of MPI-S is to notice key features of the skill as demonstrated by the mentor and ascertain the overall sense of the outcome. Students will learn how to reach the outcome in later steps of the teaching strategy.

For example, the teacher may design an investigation in high school (grades 9–12) that involves answering the question, "What are the factors that influence the height of tides on the Earth?" This investigation is an opportunity to use secondary data collected by professional scientists. Teachers can have students pull data from national data sets such as those available from the US National Oceanographic and Atmospheric Association (https://tidesandcurrents.noaa.gov/). The first part of the investigation would help students narrow their geographic focus so that they are not overwhelmed by data. One of the benefits of using a national data set is that students will gain the experience of having "messy" data, which is more realistic and authentic. Students will have to make decisions during the analysis portion of the investigation on how to clean and make choices for data timing and location so that they can answer the question.

Recall from the Next Generation Science Standards that some of the key features of analyzing and interpreting data that students in high school are expected to master are:

- Compare data sets from different groups and peer review the anomalies found between data sets
- Apply inferential statistics to answer questions or explain design solutions
- Evaluate the impact of new data on a working explanation or model

In the modeling step of the coaching strategy for this data practice, teachers can demonstrate how they would go about selecting, downloading, and cleaning the data for analysis. The teacher could first give an overview of the research question.

7.3.1.1 What Are the Factors That Influence the Height of Tides on Earth?
In the first part of the lesson, the students can develop conceptual models of what they think the factors are that influence the height of tides from their prior research. The teacher can then model how they would begin the learning task for this science and engineering practice, compare data sets, and evaluate the impact of new data on a working model. This particular question does not require inferential statistics.

The teacher should anticipate making a few mistakes in front of the class when they are modeling and explain how they recognize they made a mistake, and how they would change course to correct the mistake as best as they could. For example, the teacher can begin by downloading data that is in a time frame too short to detect the extremes of tides. The teacher can talk about how they notice that the graph is sloping up but not yet coming to a crest in order to explain how they notice that they need a larger time frame of the sample. Likewise, the teacher can download the NOAA data in increments of 6 minutes over a period of one month, which is too much data to be able to discern any patterns for the phenomenon. After visualizing the data, the teacher can demonstrate for the students how they would be more selective of the data collection. For an example of the potential downfalls of using too much data or data that is not in a period to be able to show the phenomena, see Peters-Burton et al. (2020).

By modeling the appropriate behaviors, teachers set students up for constructive goal setting and strategic planning. The teacher explained the outcome goal for the learning task, the research question, and then went about setting and planning for process goals. The process goals that the teacher demonstrated were downloading sample sets of data and visualizing them so that they could match them to the current conceptual model for data needed from a length of time (periodicity of highs and lows of tides) and amount of data to determine any patterns (every hour rather than every 6 minutes).

7.3.2 Emulation

It is during this second step of the coaching strategy that the shift from teacher-led to student-led activities begins. The emulation step is related to the SRL phase of forethought (Figure 7.1), but is different from modeling because it guides students to be the initiator of their learning about analyzing and interpreting data. During emulation, the role of the student is to replicate the science and engineering practices that the teacher models when they are given a similar task as the model. However, the student does this with considerable support, and the teacher provides the students with a checklist to set goals, and consider their self-beliefs regarding a science and engineering practice in the investigation. The checklist should begin with the outcome goal and then list the process goals that will help students define the problem they are going to pursue in the investigation.

In this example for the emulation phase, students in small groups can practice downloading data sets and comparing them for compatibility to answer the research question (outcome goal). Students are supported with a checklist of the following statements that have been addressed in the modeling step of the coaching strategy:

- I understand that I am looking for factors that influence the height of tides (outcome goal)
- I am checking my data download to be in the appropriate timeline to show patterns for tides
- I am making sure that I am not downloading too little or too much data to answer the research question
- I am using my initial conceptual model to help me keep track of the variables I am investigating
- I am using visualizations to help me make sense of the data that I am downloading
- I am keeping a list of the types of data I download so that I am not repeating the same download

Students use these statements to help them reinforce their goal setting and strategic planning for analyzing and interpreting data. Like cognitive apprenticeships, MPI-S helps students who may not have had prior access to ways of knowing in science and engineering by explicitly pointing out not just how to do the practice (procedural knowledge), but also why they are doing the practice (epistemic knowledge). In later lessons, teachers can use different checklists for different science and engineering practices.

7.4 Assessing Students in Self-Regulating Their Learning about Analyzing and Interpreting Data

7.4.1 Self-Control

The self-control step of MPI-S is related to the performance phase of SRL (Figure 7.1) because it helps students monitor their performance and focus their attention on learning about science and engineering practices. Students engaged in MPI-S to this point have observed what they are supposed to be accomplishing through the teacher model (Modeling) and have attempted similar skills and knowledge with support from the teacher (Emulation). In the third step, the teacher continues to support student self-regulation of learning about science and engineering practices but reduces support to allow students to actively reflect on their metacognitive strategies. The third and fourth steps of this coaching approach can also be used to assess how well students are beginning to self-regulate their new skills. Teachers should provide students with a more difficult attempt at the skill they are trying to build and give students only a few basic standards from which to check.

In this example, students will share the clean data sets they downloaded to determine the factors involved in the changing height of tides. Because there are so many different ways to download the data and so many different variables, students may learn more about the patterns of the phenomenon of tides and about the efficiencies of cleaning data from the other groups' experiences. Students should still be examining their conceptual models for changes in their concept of the phenomenon based on the patterns observed in their data and in other groups' data. For the science and engineering practice of analyzing and interpreting data, teachers support students by providing the following shortened checklist during their improved investigation:

- I understand the patterns that were found in my data and other groups' data during the peer review

Analyzing and Interpreting Data 125

- I am looking at differences in the data between groups and accounting for those differences
- I am making notes on the ways other groups have been efficient in cleaning and visualizing data
- I have adjusted my conceptual model based on the discussion of patterns in the data from the peer review

To check for appropriate metacognition in this step, teachers should ask a few questions about the choices that students make when they perform the practice such as:

- How did you identify the differences in the data that the groups downloaded?
- How do you know your group cleaned the data effectively?
- Did you locate portions of your initial conceptual model that can be edited or added to by the findings of the new data? Explain how you located them and how they are connected to the data.

If students struggle to answer these questions, then they likely have not mastered the learning tasks of the practice. Teachers can help students by going back and modeling the difficult practices, then giving students another chance to emulate. When students can answer these questions in a way that demonstrates their mastery of defining problems, then teachers can fade all support in the next step, self-reflection.

7.4.2   *Self-Reflection*

In the self-reflection step of MPI-S, students perform the targeted practice entirely on their own and reflect on the outcome. This last step of MPI-S is aligned to the self-reflection processes of SRL (Figure 7.1) and helps students to self-evaluate their performance and attribute their successes and failures to sources of learning. The self-reflection step builds upon the self-control step because students are expected to regulate their learning without any support. Students should be able to demonstrate they can both understand and implement the science and engineering practice (analyzing and interpreting data) without any teacher support. In this step, the teacher gives the learning task and ensures that the student is able to accomplish it in a way that parallels the science and engineering practice. The teacher can decide if the student has mastered the science and engineering practice by evaluating student answers to the questions about their rationale. Questions that a teacher would ask regarding analyzing and interpreting data for high school students engaged in the tides investigation are:

- In what ways did you notice if your downloaded data was relevant to the research question? (Goal setting and strategic planning)
- Explain how you noticed differences in the data sets during peer review. (Metacognitive monitoring)
- Explain how you identified the ways the trends in the data presented by groups or your own data changed your initial conceptual model. (Attention focusing)
- What about analyzing and interpreting data do you think you did well? Why? (Attribution)
- What about analyzing and interpreting data do you think you still need to practice? Why? (Attribution)

In the next section, there will be two case studies to demonstrate how different processes of self-regulated learning could help or hinder students in their learning while they analyze and interpret data.

7.5 Teacher–Student Cases

7.5.1 Case Study: Adaptive

The case study featured in this chapter as an example of an adaptive and effective learner for defining problems will focus on:

- Goal setting and strategic planning in the forethought phase
- Metacognitive monitoring in the performance phase
- Adaptation in the self-reflection phase

Edwin, an 11th grade student, is working with his group to answer the question, "What factors influence the height of tides on Earth?" Edwin has had success before in science and in other classes when he sets goals for himself and plans out his path to the goal. Therefore, Edwin has set the goal for himself to evaluate the impact of the new information he obtained from data to his conceptual model. He decides to plot out his strategy for getting there (strategic plan) and writes a note for himself on the checklist that the teacher gave him to write down any new idea he gets for factors related to tides and incorporate them into his model. In his investigation procedures, he makes a note in several strategic places to stop and write down the new ideas for incorporation into the models. He also writes down the research question at the top of his checklists because it helps him keep track of his end goal.

Because Edwin planned out his process goals to get to his outcome goal, he is well positioned to metacognitively monitor his progress. At every point in his procedure where he wrote down to note the new ideas he is learning into the conceptual model, he checks to be sure the ideas he writes down are answering the research question (metacognitive monitoring). He noticed that maybe he was too ambitious at the beginning of the investigation because he wrote down a reminder to write his ideas for incorporation into the conceptual model too many times. He noticed that because when he progressed to several of the reminders, he had no new information to write down. Because Edwin was metacognitively monitoring himself, he knew to erase some of the reminders that he originally wrote that did not coincide with another data download. Edwin was able to adjust his strategic plan because he was metacognitively monitoring his progress to his outcome goal.

When Edwin finished with analyzing and interpreting data, he reflected on his process of getting to his goal. He was reminded that when incorporating new information into the model, he did not have to add to the model with every move. Rather, he found that every time he analyzed a newly downloaded data set, he was able to add or adjust the conceptual model. He planned on adding his strategic notes to add to the conceptual model in the next investigation only when he had new information to analyze or a new analysis technique (adaptation).

7.5.2 Case Study: Maladaptive

The case study featured in this chapter as an example of an ineffective learner that needs support for asking questions will focus on:

- Goal setting and strategic planning in the forethought phase
- Metacognitive monitoring in the performance phase
- Adaptation in the self-reflection phase

Prior to this science class, Chester, an 11th grader in Edwin's class, was able to meet the expectations of teachers easily. However for some reason, this year Chester felt like success was out of his reach. Because science came easy to him prior to the 11th grade, he did not yet develop much of an awareness of his learning skills. As a result, Chester did not regularly and explicitly set goals for himself in science class. He knew that he and his group needed to answer the question, "What factors influence the height of tides on Earth?" but did not really understand how downloading the data as indicated in the procedure was going to help him answer that question. He followed the directions and was able to generate graphs but he really did not understand how to interpret the graphs and therefore was unaware he was going to add ideas to his conceptual model.

Although Chester was not aware of it, he set an outcome goal by knowing that he and his group needed to use data to answer the research question. However, he did not know how to strategically plan steps to get to his outcome goal (process goals). He could have used the checklists that the teacher handed out, but was distracted by the fact that he used to be "good" in science and he was now struggling since he ignored the checklists and hoped that by looking at the graphs for a long time some idea would come to him. Since he had no strategic plan and no awareness of his learning process, Chester missed opportunities to metacognitively monitor his process.

Chester did reflect on his experience during the investigation when he had finished, and his reflection caused him to be even less efficacious of his skills in science. He looked back at his experiences and felt like a failure who was completely lost and over his head. He was dreading the next time he needed to do an investigation in science and was embarrassed to ask his teacher for help.

7.5.3 *Supports the Teacher Can Provide for Chester in This Case Study*

7.5.3.1 *Goal Setting and Strategic Planning*

Chester was able to set an outcome goal because the teacher kept explicitly referring to the research question as the reason for analyzing and interpreting the data. However, Chester did not understand this to be an outcome goal of his learning. To Chester, getting an answer to the research question was just a means to getting through this investigation and he did not connect it to his learning. The teacher might have noticed that Chester had done well in science in the past and could have had a one-on-one conversation with him about his learning skills. Chester's lack of awareness of his learning skills would have been exposed then and the teacher could have explained how learning goals work and planned his strategies to meet the expectations of the investigation.

7.5.3.2 *Metacognitive Monitoring*

Chester's teacher could have used the plan for strategically reaching the outcome goal to help Chester think metacognitively through the investigation to see if he could interpret the data to incorporate it into his conceptual model. Like Edwin's streamlined plan to metacognitively monitor, the teacher could have shown Chester where to stop and interpret the data to add to the conceptual model. The teacher could also have modeled the first interpretation of data along with Chester to help him catch up with the other students in terms of being aware of his learning process.

7.5.3.3 Adaptation

At the end of the analysis and interpretation of data, Chester's teacher could have asked him to write down what worked well for him (e.g. keeping in mind the research question when he stopped to interpret data) and what did not work well for him. From the list of things that did not work well for him, the teacher could have taken some of the difficulties and offered suggestions or worked with Chester to find new ways to approach the analysis and interpretation of data next time. Although Chester felt as though he was unsuccessful, it is an emerging opportunity for him to value learning strategies that the teacher can offer. At this point in his academic career in science, Chester still cares that he is unsuccessful and the teacher can use this as motivation to become more aware of his own learning processes. If Chester can learn to be more aware of his learning skills and process, this can help him want to learn even more, and progress further in his ability to self-regulate his own learning in science.

Reflection question for teachers: Reflect on the adaptive and maladaptive case study you just read and answer the following question. What student supports for asking questions can you use in your classroom from this case study?

7.6 Designing Lessons That Use the Practice of Analyzing and Interpreting Data

In order to plan for teacher modeling of the practice "analyzing and interpreting data," teachers can review a lesson plan for ways they can model the key learning tasks for this practice in an investigation. Teachers can look for opportunities in their lesson to point out to students:

- Anomalies in different data sets
- Ways to apply inferential statistics to data
- Ways students can visualize small samples of data to better recognize patterns
- How to edit conceptual models to include new findings

Teachers can then make checklists for students by listing the key features of doing the science and engineering practice, and then convert them into student-friendly language for a checklist bullet. For example, a key feature from analyzing and interpreting data is to apply inferential statistics to answer questions or explain design solutions. The checklist bullet item for this key learning task can be "When I decided what inferential statistics to apply, I think about the type of data I have and what statistical techniques fit with that type of data." Teachers should make at least one bulleted item for each key feature of the practice. Creating more than one bulleted item could potentially help students by understanding the practice from different perspectives.

7.7 Teacher Reflection on Implementation of Lesson Featuring Analyzing and Interpreting Data

Being a reflective practitioner is important for supporting student self-regulated learning. Teachers should consider the following questions for the practice of analyzing and interpreting data:

- What worked well in supporting students in analyzing and interpreting data?
- What did not work well in supporting students in analyzing and interpreting data?
- How will I change this next time when I teach students to analyze and interpret data?

CHAPTER 8

Mathematics and Computational Thinking

The chapters about science and engineering practices (Chapters 4 through 11) have a parallel design because each of the chapters elaborates on a different science and engineering practice in detail. This chapter focuses on the practices of mathematical and computational thinking. In each chapter, the practice is dissected into distinct and clear learning tasks which can be used as process goals. These tasks are then examined within the context of a self-regulated learning cycle and points for instruction and assessment are emphasized. The tasks are reassembled into two case studies – one positive and one negative – to demonstrate how the learning tasks can be used by students.

8.1 Defining the Practice

In the Next Generation Science Standards (NGSS), mathematics and computational thinking begins with students in grades K–2 with the recognition that the world can be described mathematically. The practice of mathematics and computational thinking builds from there for students in grades 3–5 to extend quantitative measurements to a variety of physical properties and using computation and mathematics to analyze data and compare alternative design solutions. At the 6–8 grade level, the NGSS expects students to identify patterns in large data sets and use mathematical concepts to support explanations and arguments by using digital tools to find patterns and compare proposed solutions to an engineering problem, apply mathematical concepts and use mathematical representations to support scientific claims, and create algorithms to solve a problem. In grades 9–12, the NGSS builds on these ideas to introduce more mathematical functions and computational tools for statistical analysis and to model data. Students at the high school level should be able to create or revise a computational model or simulation, use mathematics and computation to design solutions, apply algebraic techniques to science and engineering problems, and use simple limit cases to test algorithms for model fit.

Using mathematics in science is mostly common in classrooms, and increasingly the science classroom has become more computational (Foster, 2006). Bringing computational tools into the classroom gives a more realistic view of these disciplines (Augustine, 2005). Computational thinking is an approach to solving problems and designing systems that requires students to think recursively, reformulate problems to see them in a different light, model relevant aspects of problems, and use abstraction and decomposition in tackling large complex problems (Wing, 2006). Computational thinking can be a useful addition to instruction as it is a suite of complex processes that students can use to become skillful in data analysis in scientific investigations (Weintrop et al., 2016).

Computational thinking can be thought of as having four components: decomposition, pattern recognition, algorithmic thinking, and abstraction. Decomposition consists of breaking the ideas into parts. Pattern recognition consists of seeing trends. Algorithmic thinking consists of using rules in new situations or being efficient with logical pathways. Abstraction consists of reducing noise to see essential elements in the ideas.

One fruitful area in a science or engineering classroom to integrate mathematics and computational thinking is in data practices. Data practices in the science classroom can be categorized into five ideas when thinking about computational thinking (Weintrop et al., 2016). Creating data occurs when students design the research methods. This can include deciding what variables to measure, what variables to keep constant, what units to use, and what tools to use. Collecting data occurs when students design their method of recording the information and the processes they use to systematically record data. Manipulating data occurs once students collect the relevant data. Students may need to clean the data (take out outliers), sort the data, or move the data around to notice patterns and explain the trends of the phenomena measured. To be scientific in manipulating data, students should be sure to follow established protocol to remove outliers or perform other manipulations. Visualizing data occurs when students decide what graph or other type of representation to use to display the data. Students need to consider how their decisions might misrepresent the data in scale or scope. Manipulating data occurs when students make conclusions or conjectures about the trends in the data. It is during this process that students make meaning from the evidence to support their claim.

In all of these data practices, students use mathematics and computational thinking to make meaningful decisions. The following matrix demonstrates how computational thinking and data practices can

Mathematics and Computational Thinking

		Computational Thinking Practices			
		Decomposition (Parts)	Pattern Recognition (Pattern)	Abstraction (Essentials)	Algorithm (Rules)
Data Practice	Creating (Identify)	Identify Parts	Identify Pattern	Identify Essentials	Identify Rules
	Collecting (Notice)	Notice Parts	Notice Pattern	Notice Essentials	Notice Rules
	Manipulating (Organize)	Organize Parts	Organize Pattern	Organize Essentials	Organize Rules
	Visualizing (Show)	Show Parts	Show Pattern	Show Essentials	Show Rules
	Analyzing (Describe)	Describe Parts	Describe Pattern	Describe Essentials	Describe Rules

Figure 8.1 Integration of computational thinking with data practices

integrate. Teachers can use computational thinking to help students perform different data practices. For example, when faced with a complex phenomenon, such as the cause of tides, students can be encouraged to use the computational thinking practice of decomposition to break down the different possible factors that cause tides. As shown in Figure 8.1, when integrating decomposition into creating data, students should be identifying the parts of the phenomena to measure for data (Peters-Burton et al., 2020).

8.2 Strategies for Teachers to Support Student Practices

For the practice of "computational thinking," we will focus on the SRL processes of *goal setting* in the Forethought phase, *attention focusing* and *metacognitive monitoring* in the Performance phase, and *self-evaluation* and *attributions* in the Self-Reflection phase as seen in Figure 8.2. We will focus on these processes because they are most illustrative for computational thinking. In other chapters discussing other practices, we will

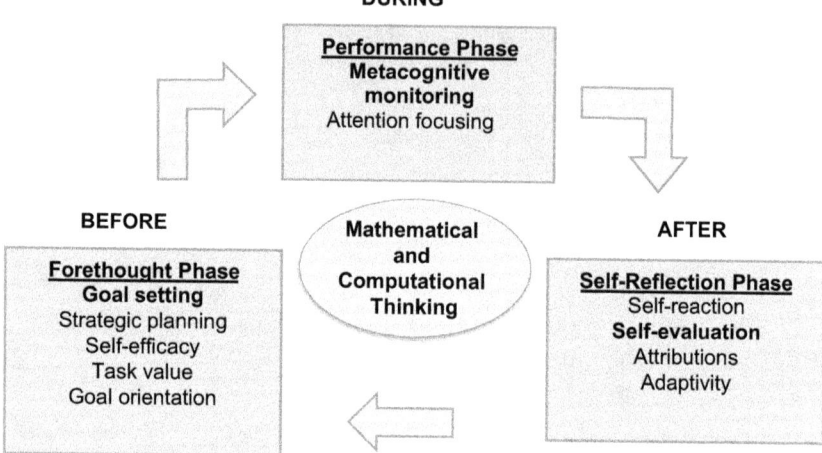

Figure 8.2 SRL processes for mathematical and computational thinking

focus on other SRL processes so that the whole cycle will be discussed across the book.

8.3 Instructing Students in Self-Regulating Their Learning about Computational Thinking

A research-based coaching strategy is available for teachers to demonstrate their strategic SRL thinking with computational thinking and data practices, called Metacognitive Promoting Intervention-Science (MPI-S; Peters, 2009; Peters & Kitsantas, 2010). MPI-S enacts SRL because it prompts students to set goals, assists students in monitoring progress toward goals, and asks students to reflect on their success in reaching those goals. This teaching strategy works well for lessons involving science and engineering practices because it engages students with the processes and approaches to thinking scientifically.

MPI-S is a suite of curricular tools, made up of a suite of checklists and questions that can be incorporated into established lesson plans to support student SRL strategies. The implementation of MPI-S consists of four steps: Modeling, Emulation, Self-Control, and Self-Regulation. The steps of MPI-S are the same ones as the coaching strategy founded by Zimmerman (2000). In the modeling phase, the teacher demonstrates mathematical and computational thinking in a lesson. The students consider their own forethought processes. In the emulation phase, the teacher provides a checklist of key learning goals in mathematical and computational thinking. The students use the checklist as a tool in a first attempt at trying the learning task. In the self-control phase, the teacher provides a short checklist and asks students questions about their processes of mathematical and computational thinking. The students monitor their progress in mathematical and computational thinking tasks. In the self-regulation phase, the teacher asks students to justify their process of using mathematical and computational thinking in an investigation. Students identify their learning instances and assess the quality of their mathematical and computational thinking.

Using this approach, the teacher initially supports students explicitly through modeling and then drops the level of support so that students are able to articulate how they understand computational thinking (or any of the science and engineering practices) independently. The first two steps of MPI-S (modeling and emulation) are instructional and the second two steps (self-control and self-regulation) assess student learning to inform instruction. MPI-S does not increase the time it takes to teach science and engineering practices, because the teaching focuses on student use of science and engineering practices that are tangible at the same time, they are creating opportunities for students to be more aware of their learning strategies.

8.3.1 Modeling

Modeling is the first step in the MPI-S teaching strategy and is aligned to forethought processes of SRL – in this case goal setting (Figure 8.2). Since students are often underexposed to the ways scientists think and conduct their work (Hogan, 2000), it is important that students begin a learning task by understanding the goal they are trying to reach. Often, students will be unable to set goals for themselves, so the teacher should step in and demonstrate what the student should be striving for. The modeling step in MPI-S helps students with their forethought processes because it demonstrates their learning goal, and helps them evaluate their self-efficacy and value in the task. Modeling is much like a cognitive apprenticeship (Collins, Brown, & Newman, 1989) where the mentor (teacher) does the activities in full view of the apprentice (students), but at the same time talks aloud about rationale, choices, and decision points with the intention that the apprentice will be able to adopt the same practices. The role of the student in this first step of MPI-S is to notice key features of the skill as demonstrated by the mentor and ascertain the overall sense of the outcome. Students will learn how to reach the outcome in later steps of the teaching strategy.

For example, the teacher may co-design an investigation that involves data practices with the class to demonstrate computational thinking. Consider the investigation that a class can do that will answer the following research question: "How does length and mass affect the time it takes a pendulum to make one full swing back and forth?" The teacher could co-create a procedure using string and washers with the students that explains how to use decomposition to create data. The teacher could draw a model of a pendulum on the board and ask students to break down and identify all possible variables involved in answering this research question (length, mass, time of swing, gravitational pull). Then the teacher could ask students to select one variable to test and collect data to pool together in a classroom spreadsheet. Traveling around the room, the teacher could talk to each group about their data collection procedures, noting patterns that emerge and the algorithm that students use to gather data. Once all student data is collected, the teacher could demonstrate manipulating data with a projected Excel spreadsheet of all of the student data. The teacher could sort accordingly and ask students to point out patterns they recognize. Then the class can decide on the type of graph they want to use to display the data using abstraction to focus on the essence of the research question. Finally, the students can analyze the visualizations and make claims about the essence of the phenomena by using abstraction.

8.3.2 *Emulation*

It is during this second step that the shift from teacher-led to student-led activities begins. The emulation step is related to the SRL phase of forethought (Figure 8.2), like modeling, but is different because it guides students to set their own goals for learning about computational thinking. During emulation, the role of the student is to replicate the scientific thinking and skills from the teacher model given a similar task as the model. However, the student does this with considerable support, and the teacher provides the students with a checklist to focus attention on a science and engineering practice in the investigation, in this case computational thinking. For example, when helping students to use computational thinking when creating data for the pendulum investigation, students are supported with a checklist of the following statements that have been addressed in the modeling step of the coaching strategy:

- I have identified all of the possible variables involved in a swinging pendulum from my model of the phenomena (decomposition)
- I have removed the variables that are not relevant to my research question (abstraction)
- I have tried to reduce the sources of interference in my procedure related to the period of a pendulum such as twisting of the washer, raising the washer to different heights, or letting the fulcrum bounce around (abstraction)
- I have designed steps for collecting data in a systematic way no matter who is collecting the data (algorithm)

Students use these statements to help them set goals that integrate computational thinking into data practices. Like cognitive apprenticeships, MPI-S helps students who may not have had prior access to ways of knowing in science by explicitly pointing out not just how to do the practice, but why they are doing the practice. In later lessons, teachers can use different checklists for different science and engineering practices.

8.4 Assessing Students in Self-Regulating Their Learning about Computational Thinking

8.4.1 Self-Control

The self-control step of MPI-S is related to the performance phase of SRL (Figure 8.2) because it helps students monitor their performance and focus their attention on learning about science and engineering practices. Students engaged in MPI-S to this point have observed what they are supposed to be accomplishing through the teacher model (Modeling) and have attempted similar skills and knowledge with support from the teacher (Emulation). In the third step, the teacher continues to support student self-regulation of learning about science and engineering practices but reduces support to allow students to actively reflect on their metacognitive strategies. The third and fourth steps of this coaching approach can also be used to assess how well students are beginning to self-regulate their new skills. Teachers should provide students with a more difficult attempt at the skill they are trying to build and give students only a few basic standards from which to check. Students are expected to take over more responsibility for learning and the teacher acts as the facilitator by proving basic support, only intervening when misconceptions arise. Using the example of the pendulum investigation in visualizing data, teachers support students by providing the following shortened checklist during their investigation:

- I have identified all of the possible visualizations that could communicate this information (decomposition)
- I have selected the type of graph that best communicates my answer to the research question (abstraction)
- I have used the appropriate scale for my visualization so as to not distort the information (pattern recognition)

To check for appropriate metacognition in this step, teachers should ask a few questions about the choices that students make when they perform the practice such as

- What are the types of graphs that could display this data?
- Why did you choose the type that you used?
- How do you know that the scale is appropriate?

If a student responds that they used a pie chart to display length versus time graph because their friend told them to, the teacher can move back into the emulation step and give more support to the student. When students can answer these questions in a way that demonstrates their mastery of computational thinking, then teachers can fade all support in the next step, self-reflection.

8.4.2 Self-Reflection

In the self-reflection step of MPI-S, students perform the targeted practice entirely on their own and reflect on the outcome. This last step of MPI-S is aligned to the self-reflection processes of SRL (Figure 8.2) and helps students to self-evaluate their performance and attribute their successes and failures to sources of learning. The self-reflection step builds upon the self-control step because students are expected to regulate their learning without any support. Students should be able to demonstrate they can both understand and implement the science and engineering practice (computational thinking) without any teacher support. In this step, the teacher gives the learning task and ensures that the student was able to accomplish it in a way that parallels scientific thinking. The teacher can decide if the student has mastered the science and engineering practice by evaluating student answers to the rationale. Questions that a teacher would ask regarding computational thinking are:

Mathematics and Computational Thinking 139

- Explain how you used decomposition in any of the data practices in the investigation. (Self-evaluation)
- How did you use an algorithm in collecting data? (Self-evaluation)
- How did pattern recognition play a part in analyzing the data? (Self-evaluation)
- What practices of computational thinking do you think you did well? Why? (Attribution)
- What computational thinking practices do you think you still need to practice? Why? (Attribution)

In the next section, there will be two case studies to demonstrate how different processes of self-regulated learning could help or hinder students in their learning of computational thinking practices.

8.5 Teacher–Student Cases

8.5.1 Case Study: Adaptive

The case study featured in this chapter as an example of an adaptive and effective learner for pattern recognition (the science and engineering practice of mathematics and computational thinking) will focus on

- Goal setting in the forethought phase
- Metacognitive monitoring in the performance phase
- Self-evaluation in the self-reflection phase

Abeer, an 8th grade student, has just collected data for an investigation that answers the question, *What is the effect of time on the temperature of ice when heated?* To collect the data, Abeer and her group started with a beaker of ice on a hot plate and took the temperature of the ice/water in the beaker every 2 minutes until the ice melted and the water began to steam. She recorded her observations in a data table. Her data table is found in Table 8.1.

She has a clear understanding of the next task in her investigation – to graph the data. She sets a goal to plot the points on a graph that has time on the x-axis and temperature on the y-axis (goal setting). As she is plotting the points, she notices that the pattern of the temperature is rising, until she gets to the 10-minute data point (pattern recognition and metacognitive monitoring). When she plots the next few points, she sees that the temperature does not rise, which she was not expecting. She asks her group members if they should go back and take the data again, but there is no time. Instead, she decides to see the other groups and if their data leveled out as well. When

Table 8.1 *Abeer's data table*

Time (min)	Temperature (Celsius)	Phase (solid, liquid, gas)	Time (min)	Temperature (Celsius)	Phase (solid, liquid, gas)
0	0	Solid	12	37	Liquid
2	4	Solid and liquid	14	40	Liquid
4	10	Liquid and solid	16	60	Liquid
6	22	Liquid	18	80	Liquid
8	37	Liquid	20	100	Liquid and gas
10	37	Liquid			

she visits other groups, she sees that the other groups had the same pattern of rising, leveling, and rising again. Abeer is now satisfied that she has recorded her observations well and when she researches why the temperature leveled off, she is able to explain her observations (self-evaluation).

8.5.2 Case Study: Maladaptive

The case study featured in this chapter as an example of an ineffective learner that needs support for decomposition (the science and engineering practice of mathematics and computational thinking) will focus on

- Self-efficacy in the forethought phase
- Attention focusing in the performance phase
- Causal attribution in the self-reflection phase

Patrick, an 8th grade student in Abeer's class, has never considered himself a good student in science, but he performed well in his English class. Patrick's teacher has heard him say several times to other students that "I'm just not a science or math person." Patrick's class was just assigned a Problem-Based Learning module asking students to design an apparatus to keep an ice cube from melting for one hour. As Patrick was setting up for the module with his new group, he was not feeling confident in his ability to do well with this project (self-efficacy). As the group got started with researching materials that can insulate from the cold, Patrick used his computer to begin playing games because he wanted to avoid science at all costs. The teacher gently redirected him to another site. Patrick stared at the site without making any notes because he was not sure what was most important (attention focusing). When Patrick received his individual grade for the group project, it was a "C." He wasn't surprised because he already felt like he wasn't science-minded and the low grade confirmed his thinking.

8.5.3 Supports the Teacher Can Provide for Patrick in This Case Study

8.5.3.1 Self-Efficacy
Patrick's low confidence for science prevents him from stretching and taking risks in learning. The teacher could help Patrick see his strong points to boost his confidence or give him an assignment in which Patrick can be successful. For a period of time, the teacher could help Patrick journal about his successes in science. For this project, the teacher could offer chunking of the tasks and a timeline (decomposition) to help Patrick feel less overwhelmed. The teacher and Patrick should work together to decompose the large task and the teacher can talk out loud about their thinking process to show Patrick how decomposition is done.

8.5.3.2 Attention Focusing
Patrick was not sure how to break a large problem down into smaller parts for solving. The teacher can ask groups to show what is their plan to solve the big task on a visualization like a Gantt chart so that they break the task down and assign smaller parts to get the larger task done. Patrick could also document when his tasks get accomplished, so that he continues to boost his self-efficacy.

8.5.3.3 Attributions
Patrick gained summative feedback from his grade, but unfortunately, because he had a hard time understanding the task, he was not able to gain feedback formatively. Because he could not clearly see what should be done for the learning task or why it should be done, he had no other information to attribute his poor grade other than his unfounded idea that he was not a science person. As the teacher helps Patrick break down the tasks in the module, she could also give Patrick a form or graphic organizer to track his progress. In attributing an external, unchangeable source for his failure (innate ability) Patrick may never really be motivated to learn science. However, if the teacher can help him see that his processes for learning lead to success or failure and help him acquire a growth mindset, Patrick may be able to discard this misconception about himself.

Reflection question for teachers: Reflect on the adaptive and maladaptive case study you just read and answer the following question. What student supports for asking questions can you use in your classroom from this case study?

8.6 Designing Lessons That Use the Practice of Using Mathematics and Computational Thinking

In order to plan for teacher modeling of the practice "using mathematics and computational thinking," teachers can review a lesson plan for ways they can model the key learning tasks in asking questions and defining problems in an investigation. Teachers can look for opportunities in their lesson to point out to students:

- Decomposition (breaking into smaller parts)
- Pattern recognition (seeing trends)
- Algorithmic thinking (applying rules)
- Abstraction (reducing noise and seeing the essence)

Teachers can then make checklists for students by listing the key features of doing the science and engineering practice, and then convert them into student-friendly language for a checklist bullet. For example, a key feature from using mathematics and computational thinking is to notice that patterns can emerge while collecting data – increasing, decreasing, and staying the same. Scientists and engineers notice if these patterns change while collecting data. The checklist bullet item for this key learning task can be "I should go back to double check data if there is a pattern that does not make sense." Teachers should make at least one bulleted item for each key feature of the practice. Making more than one item could potentially help students by understanding the practice from different perspectives.

8.7 Teacher Reflection on Implementation of Lesson Featuring Using Mathematics and Computational Thinking

Being a reflective practitioner is important for supporting student self-regulated learning. Teachers should consider the following questions for the practice of using mathematics and computational thinking:

- What worked well in supporting students' use of mathematics and computational thinking?
- What did not work well in supporting students' use of mathematics and computational thinking?
- How will I change this next time when I teach students to use mathematics and computational thinking?

CHAPTER 9

Constructing Explanations and Designing Solutions

Chapters 4 through 11 have a parallel design because each of the chapters explains how to use self-regulated learning cycles while learning with a particular science and engineering practice. This chapter is focused on "constructing explanations and designing solutions." In each chapter, the practice is dissected into distinct and clear learning tasks. These tasks are then examined within the context of a self-regulated learning cycle and a coaching strategy for instruction and assessment are explained using an example of a design challenge for middle school students related to electromagnets. The tasks are reassembled into two case studies – one positive and one negative – to demonstrate how the learning tasks can be used by students.

9.1 Learning Tasks in the Practice

The Next Generation Science Standards (NGSS Lead States, 2013) explain that the practice of constructing explanations and designing solutions focuses on the end products of an investigation. Scientists construct explanations in order to add evidence to an already established model or theory in the scientific community and to begin to develop new models that may eventually become a theory composed of several models that explain the variables, relationships, and mechanisms behind observed phenomena in the natural world. Engineers design solutions that are a product of the engineering design process that involves investigations of a design and how the design behaves given the laws of the natural world. The designed solution is expected to be efficient and effective given the constraints and resources available for the design.

These overall definitions have been further broken down into grade bands by NGSS of K–2 (aged 4 to 7), 3–5 (aged 8 to 11), 6–8 (aged 12 to 14), and 9–12 (aged 15 to 18). The following sections in the chapter explain the smaller skills that form the science and engineering practice of constructing explanations and designing solutions for the grade bands. The sections begin with the learning tasks for students from K–2 and build on the new practices that can be learned in the successive grade band (NGSS@NSTA, 2014). From there, the chapter breaks down the learning tasks involved in each skill for the practice, which represents what students should be able to do to master the practice. The chapter also provides examples of how a teacher can support the learning task in an explicit way and in a reflective way. Educational research on epistemic knowledge in science has shown that students are less likely to learn an aspect of the nature of science implicitly, but are more likely to learn that aspect when taught explicitly by the teacher and when given intentional opportunities to reflect on how the aspect was demonstrated (Khishfe & Abd-El-Khalick, 2002; Peters, 2012; Peters & Kitsantas, 2010; Peters-Burton, 2015). That literature guides practical ways to enhance student learning for science and engineering practices in this section of the chapter.

Breaking the practice down is not only important for helping students who may not have had exposure to doing science and engineering, but it is also important to situate the learning task in a cycle of self-regulated learning. In order for self-regulated learning to be productive, one must begin with a well-defined learning task. As explained in Chapter 3, goal setting is key to being successful with self-regulated learning, and the goal needs to be discrete and composed of smaller, more proximal process goals that lead up to the outcome goal.

Outcome goals in constructing explanations and designing solutions include synthesizing evidence to address the research question or design solution and connecting with already established scientific and engineering knowledge to show the role of the explanation or solution within the context of the field. However, recall that in order for students new to these practices to master them, they also need process goals that are smaller steps to reaching the outcome goals. The breakdown of characteristics for "constructing explanations and designing solutions" lays the foundation for process goals for students for this practice.

9.1.1 Process Goals for Constructing Explanations and Designing Solutions

Educational research has demonstrated that students can more effectively learn when teachers explicitly and reflectively support learning tasks (Peters-Burton, 2017). Explicit instruction does not mean that teachers should lecture about how to do the practice. Instead, teachers can model the behaviors that are shown by scientists and engineers when performing the practice of constructing explanations and designing solutions. When reflectively teaching the practice, teachers give students a chance to show what they know about performing the practice. From the assessment of the student's reflective response, teachers can determine if they need to reteach the practice or give students another chance to emulate the practice.

9.1.1.1 *Explicitly and Reflectively Supporting Process Goals for Constructing Explanations and Designing Solutions for Grades K–2*
Constructing explanations and designing solutions for students in grades K–2 builds on prior experiences and progresses to the use of evidence and ideas in constructing evidence-based accounts of natural phenomena and designing solutions (NGSS@NSTA, 2014). The following breakdown of learning tasks can be used as process goals for students in grades K–2 to learn about the practice:

- Explain natural and designed phenomena using evidence from observations
- Use tools to design a solution to a problem that has constraints
- Evaluate multiple solutions to a problem to determine the one that best fits the criteria and constraints

Teachers can explicitly model how to *explain natural and designed phenomena using evidence from observations* to show students how scientists and engineers perform the practice. As a whole class, explain the research question or design challenge and the evidence collected from observations during an investigation. Demonstrate by talking aloud about your thoughts on how you would go about explaining the phenomenon using the evidence. Pointing to the conceptual model designed at the beginning of the investigation could help with the explanation. To allow students to reflect on their skill, have students revisit their conceptual models in small groups at the end of the analysis and interpretation of data in an investigation. Have students create either written, drawn, or verbal explanations for the phenomenon that is connected to the evidence they collected in the investigation.

Teachers can demonstrate to students how to *use tools to design a solution to a problem that has constraints* by explaining the engineering design cycle to the class. Then explain that the class is about to design a solution based on the evidence collected in the "analyze and interpret data" step of the cycle. Show how you would make decisions about designing a solution based on the evidence gathered in the investigation by talking aloud about your choices, particularly about constraints. Then to facilitate student reflection, in an investigation done in small groups, have students use tools to design a solution to the problem and explain how they took into account the constraints.

Teachers can be explicit about *evaluating multiple solutions to a problem* to determine the one that best fits the criteria and constraints by presenting a design challenge to the students and then present to them multiple design solutions to the problem. Ask small groups of students to discuss the multiple design solutions and decide which fits the criteria and constraints of the design challenge in the most effective way and why. Have students present their ideas for a whole class discussion. Summarize how the designs fit the criteria and constraint and come to a decision on fit based on the whole class discussion. Then, facilitate student reflection by presenting a design challenge to the students and then present to them multiple design solutions to the problem. Ask students to have a discussion about what solution might best fit the design challenge criteria and constraints and record their thoughts on each solution.

9.1.1.2 Explicitly and Reflectively Supporting Process Goals for Constructing Explanations and Designing Solutions for Grades 3–5

According to the Next Generation Science Standards, constructing explanations and designing solutions for students in 3–5 builds on student K-2 experiences and progresses to the use of evidence in constructing explanations that specify variables that describe and predict phenomena and in designing multiple solutions to design problems. The following learning tasks can be used as process goals to support student learning to fully understand the practice:

- Construct an explanation to specify variables and relationships
- Explicitly point to evidence in the support of an explanation or design solution
- Read an explanation or design solution and point to the evidence that is supporting the explanation or solution
- Apply scientific ideas to a design problem
- Design multiple solutions for a design problem

Teachers can explicitly show students how to *construct an explanation to specify variables and relationships* so that students understand how scientists and engineers perform the practice. After data collection and analysis in an investigation, bring the class together as a group and have a discussion about how to construct an explanation while taking notes in front of the class. Demonstrate how to identify the variables and seek patterns to explain relationships. Write down notes that are the core of an explanation and demonstrate how to organize the explanation. To facilitate student reflection, after data collection and analysis in an investigation, ask student groups to construct an explanation. Ask them to note how they identified the variables and found patterns to explain relationships. Ask them to write down notes that are the key parts of their explanation and how they organized the explanation.

Teachers can help students understand how scientists and engineers *explicitly point to evidence in the support of an explanation or design* solution by modeling it for them. Present an explanation or design solution to the class that does not yet have explicit descriptions of evidence in the explanation or design solution. Demonstrate to students how to explicitly note some of the evidence in the explanation. Have student volunteers put in the rest of the evidence with your guidance. When students are constructing their explanation or designing their solution, ask them to demonstrate what they know and reflect on their skill proficiency by noting at least three pieces of evidence in their explanation or solution.

Teachers can explicitly support student learning about how to *read an explanation or design solution and point to the evidence that is supporting the explanation or solution* by presenting an explanation or design solution to the class that has explicit descriptions of evidence in the explanation or design solution. Demonstrate to students how to find some of the evidence in the explanation. Have student volunteers find the rest of the evidence with your guidance. Then to help students reflect, present an explanation or design solution to the class that has explicit descriptions of evidence in the explanation or design solution. Have students identify the evidence in the explanation and explain how they knew it was evidence.

Applying scientific ideas to a design problem can be directly modeled by the teacher to demonstrate the skill to students. After analyzing data in an investigation either as a whole class or in small groups, have the students interpret the data in a whole group with teacher guidance. Once there are patterns that are identified in the design problem investigation, have students explain why those patterns may have occurred, encouraging them to use scientific knowledge to explain the patterns. Guide them by asking what they are thinking and suggest how you would think about it. In the report of a design problem, help students demonstrate and reflect on their skill by putting a space for students to explain the patterns they see in their investigation data. Next to the spaces where the students explain patterns, put a corresponding space so that students can explain the scientific idea that is the cause of the pattern in the design solution.

Teachers can explicitly demonstrate how to *design multiple solutions for a design problem* so students understand how to go about the practice. Walk students through a design problem by using the engineering design cycle. When it comes to time for ideation, have students brainstorm multiple solutions for the design problem with your guidance. Have different groups carry out different solutions until all of the viable solutions are tested. Peer review the solutions noting the costs and benefits of each design. In a design challenge or problem, ask students to demonstrate and reflect on their skill by designing at least three different solutions for the design problem.

9.1.1.3 Explicitly and Reflectively Supporting Process Goals for Constructing Explanations and Designing Solutions for Grades 6–8
Constructing explanations and designing solutions for students in 6–8 builds on K–5 experiences and progresses to include constructing explanations and designing solutions supported by multiple sources of evidence consistent with scientific ideas, principles, and theories. The following learning tasks can be used as process goals to help students make progress to master the practices of constructing explanations and designing solutions:

- Construct an explanation or design solution using models
- Construct an explanation or design solution that uses qualitative or quantitative evidence to explain relationships between variables
- Evaluate valid and reliable evidence used to construct a scientific explanation
- Use scientific ideas or evidence to revise a scientific explanation
- Apply scientific reasoning to show why the explanation is or is not adequate for the data or evidence
- Optimize performance of a design by prioritizing criteria, making trade-offs, testing, revising, and retesting

Teachers can explicitly teach students how to *construct an explanation or design solution using models* by presenting to students a data set from an investigation, and analyze and interpret the data collaboratively. Have a discussion about what modifications are needed in the model that was created prior to the data collection. Demonstrate how to use the visualization of the model to construct the verbal explanation. When students are constructing an explanation or a design solution, have them reflect on their mastery of the skill by pointing out how their updated model matches their explanation or design solution.

Constructing Explanations and Designing Solutions

Explicit teaching of how to *construct an explanation or design solution that uses qualitative or quantitative evidence to explain relationships between variables* can help students understand how scientists and engineers approach the skill. After data collection and analysis in an investigation, bring the class together as a group and have a discussion about how to construct an explanation while taking notes in front of the class. Demonstrate how to identify the variables and seek patterns to explain qualitative and quantitative relationships. Write down notes that are the core of an explanation and demonstrate how to organize the explanation. After data collection and analysis in an investigation, ask student groups to be reflective by constructing an explanation. Ask them to note how they identified the variables and found patterns to explain qualitative and quantitative relationships. Ask them to write down notes that are the key parts of their explanation and how they organized the explanation.

Teachers can demonstrate how scientists and engineers *evaluate valid and reliable evidence used to construct a scientific explanation* by explicitly teaching the skill. Present the class with evidence that incorporates both valid evidence and unreliable evidence. Before constructing an explanation, guide students through the evaluation of the evidence – source of the evidence, sample size, and methods of gathering the data. Make notes for the class on what is valid and reliable evidence and what is not for future use. Show students how you would write an explanation or design solution using the valid and reliable evidence. Then to help students be reflective, present the class with evidence – both valid evidence and unreliable evidence. Ask students to evaluate the reliability and validity of the evidence – source of the evidence, sample size, and methods of gathering the data.

Students may have a deeper understanding of how to *use scientific ideas or evidence to revise a scientific explanation* if the teacher models it for them. Begin with an already constructed scientific explanation. Show students additional data that adds to the demonstrated explanation. Point out to students how the interpretation of the new data can be used to revise the scientific explanation. To facilitate reflective learning about the skill, give students an already constructed explanation and a data set that will add to the ideas in the explanation. Have students interpret the data and revise the scientific explanation, noting what they changed and why (have them point to the evidence they used).

Student *application of scientific reasoning to show why the explanation is or is not adequate for the data or evidence* can be modeled by teachers so that students understand what is involved. Present two explanations to the class for the same data set, one that has evidence that is adequate for the explanation and one explanation that is not adequate for the data. Demonstrate why one of the explanations needs revision by using scientific reasoning and talking aloud about your thinking. Talk aloud about your thinking and decisions for the evidence that is adequate for the explanation. Then to help students be reflective, give each student two explanations for the same data set – one that has evidence that is adequate for the explanation and one explanation that is not adequate for the data. Have the students explain, using scientific reasoning, which explanation has adequate evidence and why.

Teachers can also explicitly teach how to *optimize performance of a design by prioritizing criteria, making trade-offs, testing, revising, and retesting* so that students can see how to accomplish the skill. Present a design solution to the students and the problem it solves. List the criteria and constraints of the design problem for the students. Then present the testing results of the design solution to the students. Ask students to brainstorm how they might change the design to optimize performance. With your guidance, have them list their prioritization of criteria, costs, and benefits. As a class, make a list of the trade-offs due to the optimization. Then to help students be reflective, after students analyze and interpret data from their first set of investigation tests, ask them to redesign based on the results and on their ideas about optimization. Have them create a list of prioritized criteria and trade-offs from the revised model. Have them retest and revise once more before developing their design solution to present to clients.

9.1.1.4 Explicitly and Reflectively Supporting Process Goals for Constructing Explanations and Designing Solutions for Grades 9–12
The Next Generation Science Standards set the expectation for students in grades 9–12 to construct explanations and design solutions by building on K–8 experiences and progressing to explanations and designs that are supported by multiple and independent student-generated sources of evidence consistent with scientific ideas, principles, and theories. The following learning tasks can help these students understand the skills involved in mastering the practice:

- Construct an explanation using qualitative or quantitative data regarding the relationship and possible mechanisms
- Construct an explanation for a phenomenon using multiple sources of reliable evidence

- Design a solution and explicitly discuss taking into account possible side effects or implications

Teachers can help students understand how scientists and engineers *construct an explanation using qualitative or quantitative data regarding the relationship and possible mechanisms* by explicitly modeling it. After data collection and analysis in an investigation, bring the class together as a group and have a discussion about how to construct an explanation while taking notes in front of the class. Demonstrate how to identify the variables, seek patterns to explain qualitative and quantitative relationships, and propose possible mechanisms. Write down notes that are the core of an explanation and demonstrate how to organize the explanation. To facilitate student reflection of this skill, after data collection and analysis in an investigation, ask student groups to construct an explanation. Ask them to note how they identified the variables, found patterns to explain qualitative and quantitative relationships, and propose possible mechanisms. Ask them to write down notes that are the key parts of their explanation and how they organized the explanation.

Explicit instruction of how to *construct an explanation for a phenomenon using multiple sources of reliable evidence* can help students master the skill so that it is representative of the science and engineering disciplines. Present multiple sources of reliable evidence for the same phenomenon to students. With the students, evaluate the evidence for reliability. Demonstrate how to construct the verbal explanation using these multiple sources by writing an explanation using the first source of evidence and integrating the other sources of evidence into the narrative. To facilitate student reflection of this practice, when students are conducting an investigation, have them gather multiple sources of reliable evidence. Ask them to construct the explanation, highlighting the different parts of the explanation using different colors for different sources.

Teachers can explicitly model how to *design a solution and explicitly discuss taking into account possible side effects or implications* by presenting a design solution to the students and the problem it is solving. List the criteria and constraints of the design problem for the students. Then present the testing results of the design solution to the students. Ask students to identify any side effect or implications of the design solution based on the evidence presented. Guide the students by revealing your thinking about the evidence and any side effects and implications. After students analyze and interpret data from their first set of investigation tests, ask them to reflect on their newly learned skill by considering any side effects or implications of the solution. Have them create a list of possible side effects and/or implications and how they might reduce the impacts in the next cycle of testing.

9.2 Strategies for Teachers to Support Student Practices

For the practices of "constructing explanations and designing solutions," this section will focus on the SRL processes of *task value* in the Forethought phase, *attention focusing* in the Performance phase, and *self-reaction* and *self-evaluation* in the Self-Reflection phase as seen in Figure 9.1. The focus will be on these processes because they are most illustrative for the practices of constructing explanations and designing solutions. In other chapters discussing other practices, the book will focus on other SRL processes so that the whole cycle will be discussed across the book.

9.3 Instructing Students in Self-Regulating Their Learning about Constructing Explanations and Designing Solutions

A research-based coaching strategy has been developed for teachers to demonstrate and support strategic SRL thinking, called Metacognitive Promoting Intervention-Science (MPI-S; Peters, 2009; Peters & Kitsantas, 2010). MPI-S enacts SRL because it prompts students about setting advantageous goals for learning a task by explicitly demonstrating the practice, which gives

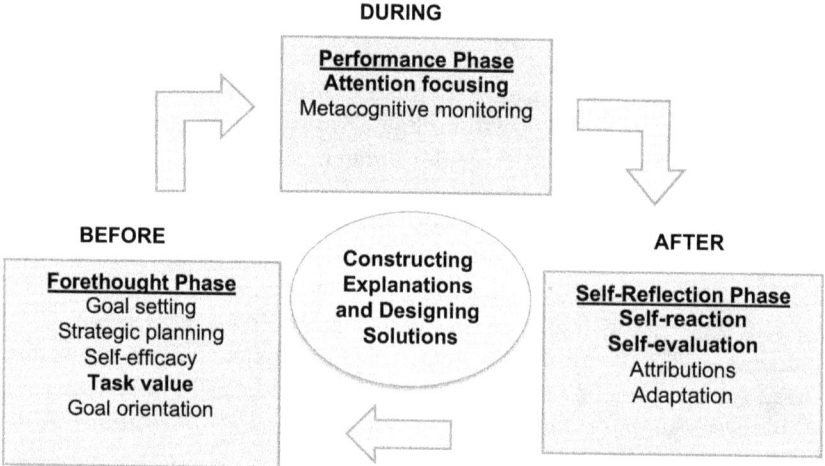

Figure 9.1 SRL processes for constructing explanations and designing solutions

students who do not have experience with the practice an entry point to learning. MPI-S also assists students in focusing their attention on the key features of planning and carrying out investigations, and asks students to reflect on their successes and failures in their performance of the practice and adapt accordingly. This teaching strategy works well for lessons involving science and engineering practices because it engages students with the processes and approaches to thinking rationally and systematically.

MPI-S is a suite of curricular tools, made up of a suite of checklists and questions that can be incorporated into established lesson plans to support student SRL strategies. The implementation of MPI-S consists of four steps: Modeling, Emulation, Self-Control, and Self-Regulation. The steps of MPI-S are the same ones as the coaching strategy founded by Zimmerman (2000). In the modeling phase, the teacher demonstrates key features of constructing explanations and designing solutions. Students consider their own forethought processes. In the emulation phase, the teacher provides a checklist of key features of constructing explanations and designing solutions. Students use the checklist as a tool for forethought processes. In the self-control phase, the teacher provides a short checklist and asks questions about student learning strategies. Students monitor their progress in constructing explanations and designing solutions. In the self-regulation phase, the teacher asks students to explain how and why they used key features of constructing explanations and designing solutions. Students identify the instances of learning and assess their quality of explanations and designing solutions.

Using this approach, the teacher initially supports students explicitly through modeling and then drops the level of support so that students are able to articulate how they understand how to construct explanations or design solutions (or any of the science and engineering practices) independently. The first two steps of MPI-S (modeling and emulation) are instructional and the second two steps (self-control and self-regulation) assess student learning to inform instruction. MPI-S does not increase the time it takes to teach science and engineering practices, because the teaching focuses on student use of science and engineering practices that are tangible and at the same time, they are creating opportunities for students to be more aware of their learning strategies.

9.3.1 Modeling

Modeling is the first step in the MPI-S teaching strategy and is aligned to forethought processes of SRL – in this case goal setting and strategic planning (Figure 9.1). Since students are often underexposed to the ways scientists think and conduct their work (Hogan, 2000), it is important that students begin a learning task by understanding the goal they are trying to reach. Often, students will be unable to set goals for themselves, so the teacher should step in and demonstrate what the student should be striving for. The modeling step in MPI-S helps students with their forethought processes because it demonstrates their learning goal, and helps them evaluate their self-efficacy and value in the task. Modeling is much like a cognitive apprenticeship (Collins, Brown, & Newman, 1989) where the mentor (teacher) does the activities in full view of the apprentice (students), but at the same time talks aloud about rationale, choices, and decision points with the intention that the apprentice will be able to adopt the same practices. The role of the student in this first step of MPI-S is to notice key features of the skill as demonstrated by the mentor and ascertain the overall sense of the outcome. Students will learn how to reach the outcome in later steps of the teaching strategy.

Recall from the Next Generation Science Standards that some of the key features of constructing explanations and designing solutions that students in middle school are expected to master are:

- Construct an explanation or design solution using models
- Construct an explanation or design solution that uses qualitative or quantitative evidence to explain relationships between variables
- Use scientific ideas or evidence to revise a scientific explanation
- Apply scientific reasoning to show why the explanation is or is not adequate for the data or evidence.

For example, the teacher may design an investigation in middle school (grades 6–8) that involves solving the design problem, *What is the optimal design for an electromagnet to pick up 25 paper clips?* Students will first work collaboratively to draw out their conceptual model of an electromagnet and identify the variables on the apparatus. For example, students can draw an iron rod with coils of wire wrapped around it and the end of the wires are attached to a power source such as a battery. From there, the students can work to identify that the number of coils, the direction of the coils, the size of the iron rod, and the strength of the power source can all be changed. Although identifying variables is a skill that moves students toward planning and carrying out investigations, it does not help students with their explanation of the results.

To get students to think about how an electromagnet works, the teacher can model their thinking about the way the variables are connected after

Constructing Explanations and Designing Solutions 155

the data is collected, analyzed, and interpreted. The teacher can refer to the outcomes of one of the variables, the size of the iron core. Using a demonstration model that all the students can see, the teacher can begin by talking about what they know about how permanent magnets work (a previous unit). The students can help the teacher think aloud about drawing the idea of domains of electrons spinning in the same direction in the iron core. Then the teacher can help students see that if there is more iron material, then there can be more domains spinning electrons in the same direction, thus increasing the magnetic force.

The teacher can also reflect on that information to see if accounting for the size of the iron core satisfies all of the connections among the variables, which it does not. The teacher can do this by looking at the data and noticing that the number of coils and the strength of the power source were also contributing to the strength of the electromagnet. The teacher can then continue to list the variables that are not yet accounted for in relationships: the coils of wire and the strength of the power source. The teacher can remind students that they learned about how when electricity travels through a wire, it causes a magnetic force from a prior unit. In the emulation phase, the teacher can guide students to explain the role of the power source and the coils of wire in the changes in magnetic force of the electromagnet.

9.3.2 Emulation

It is during this second step of the coaching strategy that the shift from teacher-led to student-led activities begins. The emulation step is related to the SRL phase of forethought (Figure 9.1), but is different from modeling because it guides students to be the initiator of their learning about constructing explanations and designing solutions. During emulation, the role of the student is to replicate the science and engineering practices that the teacher models when they are given a similar task as the model. However, the student does this with considerable support, and the teacher provides the students with a checklist to set goals, and consider their self-beliefs regarding a science and engineering practice in the investigation. The checklist should begin with the outcome goal and then list the process goals that will help students define the problem they are going to pursue in the investigation.

In the emulation phase, students can carry on the work the teacher did modeling the change in magnetic force in an electromagnet based on the size of the iron core. Students can recall that when an electric current was running through a wire, it caused a magnetic force from a prior unit. Students can then discuss in their results that when the number of coils increased, the magnetic force increased as well. Students can then incorporate that the electric current caused a magnetic force around the wire and when there was more wire, there was more magnetic force into their model. Students can also emulate the teacher's reasoning that when there was a stronger power source, there is more current generated in the wire. Students could reason that this could be the cause for increased electromagnetic strength according to the results and incorporate it into their model. Students can use the following checklist that supports some of the learning tasks.

- I understand that the goal is to increase the strength of an electromagnet to pick up twenty-five paper clips (outcome goal)
- I have referred to my initial model to identify the variables
- I have looked at my results one variable at a time and accounted for them on my model
- I have connected the results of my quantitative results one variable at a time to show relationships between variables in my model
- I have referred to ideas I have learned in previous units to provide scientific reasoning for relationships between variables
- I have gone through my data, one variable at a time, to be sure my explanation is adequate for the data

Students use these statements to help them reinforce their goal setting and strategic planning for constructing explanations and designing solutions. Like cognitive apprenticeships, MPI-S helps students who may not have had prior access to ways of knowing in science and engineering by explicitly pointing out not just how to do the practice (procedural knowledge), but also why they are doing the practice (epistemic knowledge). In later lessons, teachers can use different checklists for different science and engineering practices.

9.4 Assessing Students in Self-Regulating Their Learning about Constructing Explanations and Designing Solutions

9.4.1 Self-Control

The self-control step of MPI-S is related to the performance phase of SRL (Figure 9.1) because it helps students monitor their performance and focus their attention on learning about science and engineering practices.

Constructing Explanations and Designing Solutions 157

Students engaged in MPI-S to this point have observed what they are supposed to be accomplishing through the teacher model (Modeling) and have attempted similar skills and knowledge with support from the teacher (Emulation). In the third step, the teacher continues to support student self-regulation of learning about science and engineering practices but reduces support to allow students to actively reflect on their metacognitive strategies. The third and fourth steps of this coaching approach can also be used to assess how well students are beginning to self-regulate their new skills. Teachers should provide students with a more difficult attempt at the skill they are trying to build and give students only a few basic standards from which to check.

In this case, students should review their explanation and their revised model to see that there are three factors that influence the strength of an electromagnet: the size of the iron bar, the number of coils, and the strength of the power source. The design challenge was for students to pick up twenty-five paper clips, so the students would need to review their results to decide how large the iron bar should be, how many coils of wire it should have, and what the strength of the power source should be. The challenge for the students is to change these variables for the electromagnet to be just strong enough to pick up twenty-five paper clips, but not more or less. For the science and engineering practice of constructing explanations and designing solutions, teachers support students by providing the following shortened checklist during their improved investigation now that they have an explanation for the factors involved with the strength of an electromagnet:

- I have written explanations for the relationships among the size of the iron core, the number of wire coils, and the strength of the power source on the strength of the magnet and why those relationships occur
- The group has made trade-offs for the variables contributing to the strength of the magnet so it picks up exactly twenty-five paper clips based on our results
- We have retested the electromagnet to make sure it picks up exactly twenty-five paper clips

To check for appropriate metacognition in this step, teachers should ask a few questions about the choices that students make when they perform the practice such as:

- How did you identify the relationships between the variables in the data that the groups downloaded?

- How do you know to make changes in your model?
- How do you know your model and explanation now covers all variables and relationships related to the strength of an electromagnet?

If students struggle to answer these questions, then they likely have not mastered the learning tasks of the practice. Teachers can help students by going back and modeling the difficult practices, then giving students another chance to emulate. When students can answer these questions in a way that demonstrates their mastery of constructing explanations and designing solutions, then teachers can fade all support in the next step, self-reflection.

9.4.2 Self-Reflection

In the self-reflection step of MPI-S, students perform the targeted practice entirely on their own and reflect on the outcome. This last step of MPI-S is aligned to the self-reflection processes of SRL (Figure 9.1) and helps students to self-evaluate their performance and attribute their successes and failures to sources of learning. The self-reflection step builds upon the self-control step because students are expected to regulate their learning without any support. Students should be able to demonstrate they can both understand and implement the science and engineering practice (constructing explanations and designing solutions) without any teacher support. In this step, the teacher gives the learning task and ensures that the student was able to accomplish it in a way that parallels the science and engineering practice. The teacher can decide if the student has mastered the science and engineering practice by evaluating student answers to the questions about their rationale. Questions that a teacher would ask regarding constructing explanations and designing solutions for middle school students engaged in the electromagnet investigation are:

- In what ways did you notice if your explanation matched what you interpreted in the data? (Goal setting and strategic planning)
- Explain how you noticed patterns in the data sets that caused you to adjust your initial model of the workings of an electromagnet. (Metacognitive monitoring)
- Explain how you used your data and your model to adjust the strength of the electromagnet to pick up exactly twenty-five paper clips. (Attention focusing)
- What about constructing explanations and designing solutions you think you did well? Why? (Attribution)

- What about constructing explanations and designing solutions do you think you still need to practice? Why? (Attribution)
- What would you change about the way you construct explanations and design solutions in the future? Why? (Adaptation)

In the next section, there will be two case studies to demonstrate how different processes of self-regulated learning could help or hinder students in their learning while they construct explanations and design solutions.

9.5 Teacher–Student Cases

9.5.1 Case Study: Adaptive

The case study featured in this chapter as an example of an adaptive and effective learner for constructing explanations and designing solutions will focus on:

- Task value in the forethought phase
- Attention focusing in the performance phase
- Self-reaction and self-evaluation in the self-reflection phase

Jodie, an 8th grade student, is working with her group to construct an explanation for the factors that are involved in the strength of an electromagnet and design a solution to pick up exactly twenty-five paper clips based on their results of the investigation. Jodie was mildly interested in science at the beginning of the school year, but the way her teacher presented science as a way to wonder about the world rather than memorize facts has stoked her interest in the natural world. Jodie had never really thought about electromagnets before her teacher gave her the learning task, but after she started learning about the relationship between electricity and magnetism, she started seeing examples of electromagnets in her life. For example, Jodie recognized the electromagnet involved when she opened the trunk of her mom's car. Jodie began to value the task of understanding the relationships of the variables in electromagnets when she started tinkering with some old speakers that were in her garage. Jodie generally wanted to learn about the world around her and had a high task value for the learning task as a result.

Now that Jodie was seeing electromagnets around her in her daily life, she could see the need for them but did not yet have the skills to start to understand all of the variables involved and how they connected. When her teacher modeled using the data to update the conceptual model the students built of an electromagnet, Jodie began to see how the variables connected (attention focusing). Jodie used the checklist that her teacher gave her during the emulation and self-control phases to move beyond a general idea of electromagnets in application to see the scientific principles involved (attention focusing). When she saw electromagnets in her daily life, she would explore and try to locate the core, the coils, and the power source in the examples that she saw. Teacher modeling and her repeated exercise of looking at electromagnets in different contexts solidified her understanding of the variables and relationships involved in the phenomenon.

When Jodie constructed her explanation of the relationships between variables in an electromagnet, she received feedback from her peers and her teacher. She had connected how the size of the iron core and the strength of the power source were related to the strength of an electromagnet well, according to the feedback she received. She was pleased with her performance (self-reaction). However, the feedback she received showed that she was not quite connecting the number of coils with the strength of the electromagnet, and had represented the length of the wire rather than number of coils to the electromagnet strength. When she received feedback that she was on the right track with the idea that a longer wire produced more magnetic field, but that the magnetic field had to be positioned around the iron core to influence the domains, she was at first taken aback. She was disappointed in herself that she did not understand the model perfectly (self-reaction). However, after a moment, she regrouped and shifted her paradigm, realizing the wire and the iron core worked together. She grasped that it was a small shift and she felt that her performance was still a good one (self-evaluation), remembering that she needed to double-check her model with her data after thinking about the relationships in the future (adaptation).

9.5.2 Case Study: Maladaptive

The case study featured in this chapter as an example of an ineffective learner that needs support for asking questions will focus on:

- Task value in the forethought phase
- Attention focusing in the performance phase
- Self-reaction and self-evaluation in the self-reflection phase

Giuseppina, an 8th grader in Jodie's class, was interested in science, but was focused only on biology and animal behavior. When she approached the electromagnet task, she did not value the task because she was really only interested in biology and animal behavior (lack of task value). She felt that her time was better spent dog sitting for her aunt instead of thinking about electromagnets. As a result, Giuseppina had low task value for the learning task and decided to spend her time on other things.

Giuseppina just wanted to get through the school day so she could check on her aunt's dog, and therefore did not spend much effort in paying attention to the ways that she was constructing explanations and designing solutions. She checked items on her supporting checklists during the emulation and self-control phases of the investigation if she could understand the sentence, rather than checking to see if she could really do what the checklist was suggesting (lack of attention focusing). When her group asked her how she could have checked off items on the checklist that they have not performed yet, she bluffed by telling them she did this stuff in science camp last summer.

When Giuseppina received feedback on her model, explanation, and design solution for the electromagnet, it was unsatisfactory. She looked at the notes on her work, turned it over, and vowed never to think about this investigation again (self-reaction). She figured that she would not need knowledge about physical science in her future career as an animal behaviorist (self-evaluation) and thought about how she could keep her class grade high enough to pass the class.

9.5.3 Supports the Teacher Can Provide for Giuseppina in This Case Study

9.5.3.1 Task Value
The teacher knew about Giuseppina's love of animals and their behavior, and tried to get Giuseppina to understand that physical science is related to her understanding of the natural world. Meanwhile, Giuseppina was unconvinced that machines and motors were to be valued in the same way as animals. The teacher could then further engage Giuseppina in electromagnetic phenomena by connecting them to animals. The teacher could have introduced Giuseppina to a few developmentally appropriate articles about the way low levels of electromagnetic noise were disrupting migratory birds' compasses (Yong, 2014). By having a quick one-on-one conversation at the beginning of the investigation about the articles, the teacher could have improved Giuseppina's task value and helped her take learning risks and persist when she struggled.

9.5.3.2 Attention Focusing
Since the teacher might not have the time to stand by Giuseppina's side and help her focus her attention on the key characteristics of the skills for

this science and engineering practice, she could have paired Giuseppina with a student that she trusts and one that will use the checklists actively. For example, the teacher could have paired Jodie and Giuseppina in the same group because the teacher knew they would get along well and that Jodie could continue to model the use of the checklists to succeed in meeting her process goals on the way to the outcome goal. The teacher could optimize their time by checking on them as a pair and seeing if they were both using the checklists to emulate the skill of adding concepts to the model from the results of the investigation to create an explanation.

9.5.3.3 Self-Reaction and Self-Evaluation

When Giuseppina received feedback on her work, she noticed that she did not yet meet the expectations of the learning task and ignored the opportunity to reflect on her work. When the teacher saw her turn her paper over, the teacher could have made a mental note to discuss this with her in a one-on-one situation in the near future. When Giuseppina and the teacher meet, the teacher could talk to Giuseppina about her motivations in not wanting to learn from her mistakes. The teacher could give examples of how the teacher has personally learned from their mistakes and give examples of how the teacher sets up the class to embrace making mistakes and learning from them. Teacher intervention on Giuseppina's task value, attention focusing, self-reaction, and self-evaluation could be a powerful set of strategies to help Giuseppina to be more aware of her learning skills in the future.

Reflection question for teachers: Reflect on the adaptive and maladaptive case study you just read and answer the following question. What student supports for asking questions can you use in your classroom from these case studies?

9.6 Designing Lessons That Use the Practice of Constructing Explanations and Developing Solutions

In order to plan for teacher modeling of the practice "constructing explanations and developing solutions," teachers can review a lesson plan for ways they can model the key learning tasks in asking questions and defining problems in an investigation. Teachers can look for opportunities in their lesson to point out to students:

- How to identify variables on a conceptual model
- Ways to apply interpretation of patterns in data to adaptations in a conceptual model when questions are testable and scientific

- Ways to revise models based on results of data
- How to use data to optimize the outcome by applying trade-offs of variable influences

Teachers can then make checklists for students by listing the key features of doing the science and engineering practice, and then convert them into student-friendly language for a checklist bullet. For example, a key feature from constructing explanations and designing solutions is to construct an explanation or design solution that uses qualitative or quantitative evidence to explain relationships between variables. The checklist bullet item for this key learning task can be "When writing an explanation, I look at the model I revised from the data to help me write the connections between variables." Teachers should make at least one bulleted item for each key feature of the practice. Making more than one item could potentially help students by understanding the practice from different perspectives.

9.7 Teacher Reflection on Implementation of Lesson Featuring Constructing Explanations and Designing Solutions

Being a reflective practitioner is important for supporting student self-regulated learning. Teachers should consider the following questions for the practice of constructing explanations and designing solutions:

- What worked well in supporting students to construct explanations and design solutions?
- What did not work well in supporting students to construct explanations and design solutions?
- How will I change this the next time I teach students to construct explanations and design solutions?

CHAPTER 10

Engaging in Argument from Evidence

Each of the chapters in Chapters 4 through 11 has a parallel design that discusses a particular science and engineering practice in detail. This chapter focuses on the practice of "engaging in argument from evidence." In each chapter, the practice is dissected into distinct and clear learning tasks that can be used as process goals. These tasks are then examined within the context of a self-regulated learning cycle. A multistep coaching strategy explaining points for instruction and assessment is provided using the example of a design challenge for students in grades K through 2. The design challenge asks students to identify a problem to solve from a picture book and use the engineering design process to solve that problem. The tasks (process goals) are reassembled into two case studies – one positive and one negative – to demonstrate how the learning tasks can be used by students.

10.1 Learning Tasks in the Practice

The Next Generation Science Standards (NGSS Lead States, 2013) note that the practice of engaging in argument from evidence is done by both scientists and engineers. They use argumentation to determine the most logical and rational explanation for a phenomenon or the design solution that addresses the problem and fits within the criteria and constraints. Argumentation in science and engineering consists of the use of evidence and reasoning to demonstrate the validity of a claim. Scientists can use argumentation to present new ideas given new evidence and engineers can use argumentation to present evidence for a particular idea for a design solution. The strength of an argument is dependent on the strength of evidence, the connection to the evidence, and how well the reasoning connects the evidence and the claim.

These overall definitions have been further broken down into grade bands by NGSS of K–2 (aged 4 to 7), 3–5 (aged 8 to 11), 6–8 (aged 12 to 14), and 9–12 (aged 15 to 18). The following sections in the chapter explain smaller learning tasks that lead a student to master the science and engineering practice of engaging in argument with evidence for the grade bands. The sections in the chapter begin with the learning tasks from K–2 and build on the new practices that can be learned in the successive grade bands (NGSS@NSTA, 2014). The sections also provide examples of how a teacher can support the learning task in an explicit way and in a reflective way. Educational research on epistemic knowledge in science has shown that students are less likely to learn an aspect of the nature of science implicitly, but are more likely to learn that aspect when taught explicitly by the teacher and are given intentional opportunities to reflect on how the aspect was demonstrated (Khishfe & Abd-El-Khalick, 2002; Peters, 2012; Peters & Kitsantas, 2010; Peters-Burton, 2015). That literature provides ideas for practical ways to enhance student learning for science and engineering practices.

Breaking the practice down is not only important for helping students who may not have had exposure to doing science and engineering, but it is also important to situate the learning task in a cycle of self-regulated learning. In order for self-regulated learning to be productive, one must begin with a well-defined learning task. As explained in Chapter 3, goal setting is key to being successful with self-regulated learning, and the goal needs to be discrete and composed of smaller, more proximal process goals that lead up to the outcome goal.

Outcome goals for engaging in argument using evidence include stating a claim that is closely tied to the evidence, using reasoning to connect evidence and a claim, using multiple sources of evidence to make a claim, and evaluating arguments made by others. However, recall that in order for students new to these practices to master them, they also need process goals that are smaller steps to reaching the outcome goals. The key characteristics for "engaging in argument using evidence" can lay the foundation for process goals for students.

10.1.1 Process Goals for Engaging in Argument Using Evidence

Educational research has demonstrated that students can more effectively learn when teachers explicitly and reflectively support learning tasks (Peters-Burton, 2017). Keep in mind that explicit teaching does not mean lecturing. Instead, teachers can model the behaviors that are demonstrated by scientists and engineers when performing the practice of engaging in argument using evidence when explicitly teaching the practice. When reflectively teaching the practice, teachers give students a chance to show what they know about performing the practice. From assessing the student's reflective response, teachers can determine if they need to reteach the practice or give students another chance to emulate the practice.

10.1.1.1 Explicitly and Reflectively Supporting Process Goals for Engaging in Argument Using Evidence for Grades K–2

According to the Next Generation Science Standards, engaging in argument from evidence for students in K–2 builds on prior experiences and progresses to comparing ideas and representations about the natural and designed worlds (NGSS@NSTA, 2014). The following breakdown of learning tasks can be used as process goals for students to understand how to master the practice:

- Make a claim and support it with evidence
- Listen to an argument and identify the main points of an argument
- Listen to an argument and indicate agreement or disagreement using evidence
- Identify arguments that are supported by evidence
- Distinguish between evidence that is related to a claim and evidence that is not
- Distinguish between opinion and evidence in an argument

Teachers can support K–2 students explicitly by *modeling how to make a claim and support it with evidence*. Ask students what they wonder about regarding the natural world. Take one of the topics of interest and make a claim, such as plants need sunlight to thrive. Demonstrate to students how you would go about collecting evidence, such as placing a few plants on a classroom windowsill, some in the closet, and some in the corner of the room that has only indirect light. After a period of time, make the claim again and show students the results of the plant growth as evidence. To help students to be reflective, after students conduct an investigation, ask them to summarize what they learned by making a claim. Ask them "How do you know that? What evidence do you have?"

Students can learn more about the practice of *engaging in argument using evidence* by observing teachers modeling how to listen to an argument and identify the main points of an argument. Highlight the claim and evidence that supports the claim explaining how you identified the claim and how you identified the evidence. To support student reflection, explain a full argument about a phenomenon or design solution to students. Ask them to explain the main points of the argument.

Engaging in Argument from Evidence 167

Teachers can model for students how to *listen to an argument and indicate agreement or disagreement using evidence*. Present a full argument to students about a phenomenon or design solution. Highlight the claim and evidence that supports the claim. Explain whether you agree or disagree with the argument and why. Make explicit connections to the evidence in your reasoning and show students that you must explicitly connect to the evidence in evaluating an argument. Teachers can support students' reflection about the skill by telling students a full argument about a phenomenon or design solution. Ask them to agree or disagree with the argument and explain their reason for their decision using evidence.

Teachers can explicitly show students how to *identify arguments that are supported by evidence* in order to help students see how scientists and engineers perform the practice. Present to students a collection of arguments, some that are supported by evidence and some that are not supported by evidence. Sort the arguments into evidence and no evidence, explaining your thinking aloud. Make a checklist for students that helps them identify evidence in an argument. To support student reflection for this skill, present to students a collection of arguments, some that are supported by evidence and some that are not supported by evidence. Allow them to use the checklist to sort the arguments into ones that have evidence and ones that do not have evidence.

Teachers can model how to *distinguish between evidence that is related to a claim and evidence that is not related to a claim* to help students see how engineers and scientists perform the practice. Present students with a claim and a list of evidence for the claim, some that is related to the claim and some that is not. Walk the students through the list, identifying what makes the evidence related. With help from the whole class, make a list of how to identify evidence that is related to a claim and how to identify evidence that is not related so it can be set aside. Add key characteristics for what makes evidence related to the checklist used to identify evidence. To support student reflection, present students with a claim and a list of evidence for the claim, some that is related to the claim and some that is not. Ask students to go through the list and identify evidence that is related to the claim and evidence that is not and why they think so.

Students who can *distinguish between opinion and evidence in an argument* are more proficient in the practice, and teachers can support this skill by explicitly teaching it. Present students with a claim and a list of evidence and opinions for the claim. Walk the students through the list, identifying evidence and opinion, talking aloud about your reasoning. With help from the whole class, make a list of how to identify evidence and how to identify opinions so they can be set aside in an argument. Add key characteristics for how to exclude opinions to the checklist used to identify evidence. To support students reflectively, present students with a claim and a list of evidence and opinions for the claim. Ask students to go through the list and identify evidence and identify opinions and why they have identified them as such.

10.1.1.2 Explicitly and Reflectively Supporting Process Goals for Engaging in Argument Using Evidence for Grades 3–5
According to the Next Generation Science Standards, engaging in argument from evidence for students in grades 3–5 builds on K–2 experiences and progresses to critiquing the scientific explanations or solutions proposed by peers by citing relevant evidence about the natural and designed worlds. The following learning tasks can be used as process goals to help students in grades 3–5 understand what is involved in engaging in argument from evidence:

- Distinguish between reasoned judgment and speculation in an argument
- Evaluate an argument based on the presented evidence
- Refine an argument by making stronger connections to evidence
- Create an argument for the strength or weakness of an investigation procedure, model, or explanation
- Make a claim about a design solution based on how well the evidence aligns to the criteria and constraints

Teachers can explicitly help students understand how to *distinguish between reasoned judgment and speculation in an argument* by modeling the behavior. Present students with arguments, some that have reasoned judgment and some that have speculation. Walk the students through the list, identifying those with reasoned judgment and those with speculation, talking aloud about your reasoning. With help from the whole class, make a list of how to identify reasoned judgment and speculation in an argument. To help students reflect on how well they have learned the skill, present students with arguments, some that have reasoned judgment and some that have speculation. Ask students to go through the list and identify those that have reasoned judgment and those that have speculation and why they have identified them as such.

Teachers can demonstrate to students how to *evaluate an argument based on the presented evidence* so that students can later emulate the behavior. Present to students a claim and evidence that has some misalignments or

Engaging in Argument from Evidence 169

some poor evidence. Talk about the evidence, one piece at a time, to the students and evaluate the alignment, the completeness of the evidence, and the reliability of the evidence. Summarize your evaluation of the argument, talking aloud about your choices. Teachers can support student reflection on their skills by giving the students a claim and evidence to evaluate with a graphic organizer with space so that students can evaluate each piece of evidence and the claim.

Teachers can model how to *refine an argument by making stronger connections to evidence* for students to explicitly understand the underpinnings of the skill. Present to students a claim and evidence that has some misalignments or some poor evidence. Talk about the evidence, one piece at a time, to the students and evaluate the alignment, the completeness of the evidence, and the reliability of the evidence. Demonstrate to students how you can obtain more or different evidence to make the connections to the claim stronger, talking aloud about your thinking. To help students reflect on the proficiency of their skill, give the students a claim and evidence to evaluate with a graphic organizer with space so that students can evaluate each piece of evidence and the claim. Once students have evaluated the argument, ask them to report on how to collect more or different evidence to make a stronger argument. Ask them to provide their reasoning for their choices.

Modeling how to *create an argument for the strength or weakness of an investigation procedure, model, or explanation* can help students see how to go about this task as a scientist or engineer would. After developing a procedure for an investigation as a class, evaluate the procedure and create a list of the pros and cons of the procedure. Use the pro/con list to create an argument for the strength of the procedure, clearly indicating the claim you are making and the evidence you are using to back it up from the pro/con list. After students develop procedures for an investigation, have them reflect by peer reviewing another group's procedure. Have students create an argument for the strength of the procedure, indicating the claim they are making and the evidence they are using to back it up.

Teachers can also model for students how to *make a claim about a design solution based on how well the evidence aligns to the criteria and constraints*. Present to students a design challenge that includes criteria and constraints, the design solution, and the evidence produced from the tests of the design. Gather evidence about the alignment to the criteria and constraints on a list or graphic organizer along with the whole class. Use this evidence to make a claim, talking aloud about the decisions you make to create the claim. Then, help students to reflect so that they can self-evaluate their skills. After students complete a design challenge, ask them to make an argument for their design solution based on how well the solution aligns to the criteria and constraints.

10.1.1.3 Explicitly and Reflectively Supporting Process Goals for Engaging in Arguments with Evidence for Grades 6–8

The practice of engaging in arguments with evidence for students in grades 6–8 is defined by the Next Generation Science Standards as building on K–5 experiences and progresses to constructing a convincing argument that supports or refutes claims for either explanations or solutions about the natural and designed worlds. The following learning tasks can be used as process goals for students to break down the underpinnings of the practice:

- Use reasoning to explain why evidence supports a claim
- Analyze two arguments on the same topic to determine their strength
- Construct both oral and written arguments that include a claim, evidence, and reasoning
- Use an argument to critique a scientific explanation or design solution

Teachers can model how to *use reasoning to explain why evidence supports a claim* for students. Present an argument and with help from the class, decompose the argument into claim and evidence. Demonstrate that reasoning is also a part of an argument and it is used to connect the evidence to the claims by showing how the results of the evidence would work to make the claim. To help students reflect on their performance of the practice, give students an argument with only a claim and supporting evidence. Ask students to provide reasoning (scientific principles) to explain how the evidence supports the claim.

Teachers can help students by explicitly teaching how to *analyze two arguments on the same topic to determine their strength*. Create a graphic organizer that can be used to decompose an argument into claim, evidence, and reasoning. Use this graphic organizer in front of the class to decompose two arguments on the same topic. Walk students through an evaluation of the strength of the evidence provided for each argument, the connection of evidence to the claim, and the ability of the reasoning to connect evidence and claim while talking aloud about your choices. Summarize your thoughts on both arguments and demonstrate to students your thoughts on the strength of each argument. Teachers can help students reflect on their performance of the skill by providing students with the graphic organizer to decompose both arguments into claim, evidence, and reasoning. Ask students to evaluate the strength of evidence, the clarity of the claim, and the ability of the reasoning to connect claim and evidence. Ask students to explain which argument they think is the stronger one and why.

Teachers can model how to *construct both oral and written arguments that include a claim, evidence, and reasoning* so that students know exactly what is expected in order to behave like a scientist or engineer. Provide students with data that has been interpreted from an investigation, along with research questions, the conceptual model, and methods of collecting the data. Model to students how to construct a written argument that makes a claim from the data interpretation, backs up the claim with evidence from the data, and provides reasoning to connect the claims to the evidence. For an oral argument, create a rubric with the help of the class that lists the characteristics of a good oral argument with levels of proficiency to master. Teachers can also help students evaluate their progress in learning the skill by asking them to perform the skill and reflect. After the data interpretation portion of an investigation, have students create a written argument that summarizes their findings in the form of a claim, using evidence from the investigation to support the claim and reasoning to connect the evidence to the claim. For an oral argument, conduct a class argument to support a particular claim. Have students do research to find evidence and reasoning to support or refute the claim. Students in the class can use the rubric to guide their verbal argument in the class discussion.

Teachers can support student skill development by modeling how to *use an argument to critique a scientific explanation or design solution*. Present a scientific explanation or design solution to the class. Demonstrate how you would go about critiquing the explanation or design solution in a systematic way. For example, going line by line through the evidence to see if it supports the claim, is aligned to the problem or question posed, and is empirically strong. Then look for any missing evidence or reasoning. Finally, critique the wording of the claim. To help students to perform and reflect on the skill, give students a scientific explanation and a conceptual model. Ask students to conduct research to find evidence to critique the explanation and model. Ask students to explain their process for critiquing the explanation.

10.1.1.4 Explicitly and Reflectively Supporting Process Goals for Engaging in Argument Using Evidence for Grades 9–12

According to the Next Generation Science Standards, engaging in argument from evidence for students in 9–12 builds on K–8 experiences and progresses to using appropriate and sufficient evidence and scientific reasoning to defend and critique claims and explanations about the natural and designed worlds. Arguments may also come from current scientific or historical episodes in science. The following learning tasks can be used as process goals so that students understand the underpinnings of the practice:

- Create a counterargument as a way of rebutting another person's claim
- Evaluate the merits of two competing arguments and make an explicit argument for why one argument is stronger and why the other is weaker
- Compare two competing arguments, and construct a new argument which justifies why it is superior to previous arguments
- Use an argument to justify changes to a conceptual model

Teachers can explicitly teach students how to *create a counterargument as a way of rebutting another person's claim* so that students can understand what is involved. Explain to the class the argument made by Lamarck regarding the theory of evolution. Break down Lamarck's claim, evidence, and reasoning for the class. Then demonstrate some of the evidence that was collected by Darwin with regard to natural selection. Systematically create a counterargument as a way of rebutting Lamarck's theory, all the while talking aloud about your thinking. To assist students in reflecting on the performance of their skill, give the students an argument for a geocentric model of the Earth, including the claim, evidence, and reasoning. Ask them to create a counterargument for a heliocentric model of the Earth, using evidence found in historical research.

Teachers can help students to learn how to *evaluate the merits of two competing arguments and make an explicit argument for why one argument is stronger and why the other is weaker* by modeling it for them. Create a graphic organizer that can be used to decompose an argument into claim, evidence, and reasoning. Use this graphic organizer in front of the class to decompose two arguments on the same topic. Walk students through an evaluation of the strength of the evidence provided for each argument, the connection of evidence to the claim, and the ability of the reasoning to connect evidence and claim while talking aloud about your choices. Summarize your thoughts on both arguments and demonstrate to students your thoughts on the strength of each argument. Then, teachers can ask students to reflect on their performance by providing students with the graphic organizer to decompose both arguments into claim, evidence, and reasoning. Ask students to evaluate the strength of evidence, the clarity of the claim, and the ability of the reasoning to connect claim and evidence. Ask students to explain which argument they think is the stronger one and why.

Teachers can model how they go about *comparing two competing arguments, and constructing a new argument which justifies why it is superior to previous arguments* so that students can see some strategies for accomplishing this goal. As a whole class decompose two arguments on the same topic. Walk students through an evaluation of the strength of the evidence provided for each argument, the connection of evidence to the claim, and the ability of the reasoning to connect evidence and claim while talking aloud about your choices. Summarize your thoughts on both arguments and demonstrate to students your thoughts on the strength of each argument. From your summarization create a new claim, indicating the evidence and claims that are relevant. Teachers can help students reflect on their performance of the skill by providing students with the graphic organizer to decompose both arguments into claim, evidence, and reasoning. Ask students to evaluate the strength of evidence, the clarity of the claim, and the ability of the reasoning to connect claim and evidence. Ask students to create a new argument and justify why it is superior to the prior arguments.

Teachers can explicitly teach how students can *use an argument to justify changes to a conceptual model* so that it aligns to science and engineering disciplinary standards. Present an initial conceptual model to the class and then present data that has been interpreted from a new investigation related to the model. Identify the new information found from the interpreted data and incorporate it into the model. Identify evidence that is new from the interpreted data and make a claim in light of the new evidence in front of the class while talking aloud about your thinking. Use scientific reasoning to connect the new evidence to the new claim to justify the changes in the conceptual model. Then, teachers can reflectively teach the skill by asking students to perform the skill on their own. During an investigation, when students interpret data and make changes to their conceptual model, ask them to make an oral or written argument that justifies the changes in the conceptual model. Students should use the interpreted data from the investigation as evidence for their argument.

10.2 Strategies for Teachers to Support Student Practices

For the practices of "engaging in arguments using evidence," this section will focus on the SRL processes of *self-efficacy* in the Forethought phase,

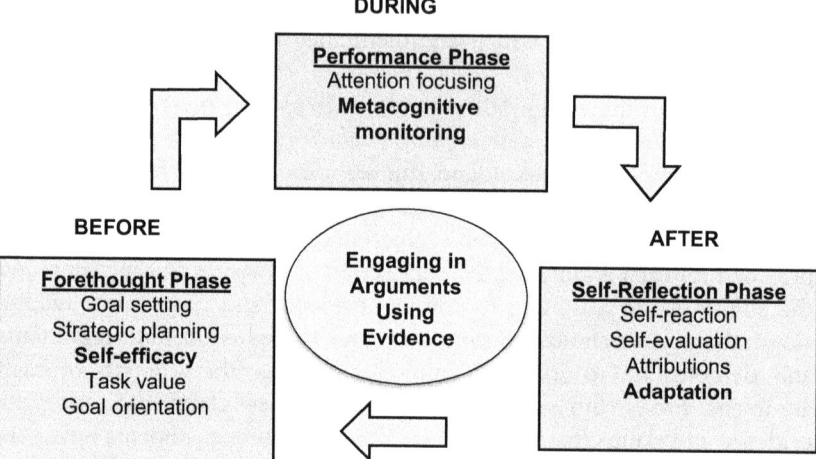

Figure 10.1 SRL processes for engaging in arguments using evidence

metacognitive monitoring in the Performance phase, and *adaptation* in the Self-Reflection phase as seen in Figure 10.1. The focus will be on these processes because they are most illustrative for the practices of engaging in argument using evidence. In other chapters discussing other practices, the book will focus on other SRL processes so that the whole cycle will be discussed across the book.

10.3 Instructing Students in Self-Regulating Their Learning about Engaging in Arguments from Evidence

The Metacognitive Promoting Intervention-Science (MPI-S) is based on a research-based teaching strategy that is used to help teachers support SRL for students (Peters, 2009; Peters & Kitsantas, 2010). MPI-S enacts SRL because it prompts students about setting advantageous goals for learning a task by explicitly demonstrating the practice, which gives students who do not have experience with the practice an entry point to learning. MPI-S also assists students in focusing their attention on the key features of engaging in argument using evidence, and asks students to reflect on their successes and failures in their performance of the practice and adapt accordingly. This teaching strategy works well for lessons involving science and engineering practices because it engages students with the processes and approaches to thinking rationally and systematically.

MPI-S is a suite of curricular tools, made up of a suite of checklists and questions that can be incorporated into established lesson plans to support student SRL strategies. The implementation of MPI-S consists of four steps: Modeling, Emulation, Self-Control, and Self-Regulation. The steps of MPI-S are the same ones as the coaching strategy founded by Zimmerman (2000). In the modeling phase, the teacher demonstrates key features of engaging in arguments using evidence. Students consider their own forethought processes. In the emulation phase, the teacher provides a checklist of key features of engaging in arguments using evidence. Students use the checklist as a tool for forethought processes. In the self-control phase, the teacher provides a short checklist and asks questions about student learning strategies. Students monitor their progress for engaging in arguments using evidence. In the self-regulation phase, the teacher asks students to explain how and why they used key features of engaging in arguments using evidence. Students identify the instances of learning and assess their quality of engaging in arguments using evidence.

Using this approach, the teacher initially supports students explicitly through modeling and then drops the level of support so that students are able to articulate how they understand how to engage in arguments using evidence (or any of the science and engineering practices) independently. The first two steps of MPI-S (modeling and emulation) are instructional and the second two steps (self-control and self-regulation) assess student learning to inform instruction. MPI-S does not increase the time it takes to teach science and engineering practices, because the teaching focuses on student use of science and engineering practices that are tangible at the same time, they are creating opportunities for students to be more aware of their learning strategies.

10.3.1 Modeling

Modeling is the first step in the MPI-S teaching strategy and is aligned to forethought processes of SRL – in this case goal setting and strategic planning (Figure 10.1). Students are often underexposed to the ways scientists think and conduct their work (Hogan, 2000), therefore they may not have high self-efficacy to accomplish the practices of engaging in arguments using evidence. Helping students to improve their self-efficacy for engaging in arguments using evidence can help them take more academic risks, which can result in student persistence in the face of failure. High self-efficacy also helps students to challenge themselves and stretch to meet goals that improve the way they perform a science and engineering practice. On the other hand, low self-efficacy can cause students to avoid mastery because they are either afraid to try or are not willing to persist in their learning. According to Bandura (2002), there are four sources of self-efficacy, which are listed from most influential to least influential:

- Having performed a similar task and been successful
- Having witnessed a peer or someone they trust perform a similar task successfully
- Being encouraged by a mentor
- Being excited (or conversely being anxious) about taking on the task

The modeling step in MPI-S helps students with their forethought processes such as self-efficacy because it demonstrates their learning goal, and helps them evaluate their self-efficacy and value in the task. In this case students may realize through teacher modeling that they have performed a similar task and been successful or by witnessing the modeling, they can visualize their success. Alternatively, students may witness the modeling and become overwhelmed. Therefore, teachers should take care to break down the key features of the practice of asking questions and defining problems so that students can set smaller goals and feel higher self-efficacy. Teachers can also encourage students doing the practice during later phases of the coaching strategy.

Modeling is much like a cognitive apprenticeship (Collins, Brown, & Newman, 1989) where the mentor (teacher) does the activities in full view of the apprentice (students), but at the same time talks aloud about rationale, choices, and decision points with the intention that the apprentice will be able to adopt the same practices. The role of the student in this first step of MPI-S is to notice key features of the skill as demonstrated by the mentor and ascertain the overall sense of the outcome. Students will learn how to reach the outcome in later steps of the teaching strategy.

Recall from the Next Generation Science Standards that some of the key features of constructing explanations and designing solutions that students in kindergarten through second grade are expected to master are:

- Make a claim and support it with evidence
- Listen to an argument and identify the main points of an argument
- Listen to an argument and indicate agreement or disagreement using evidence
- Identify arguments that are supported by evidence
- Distinguish between evidence that is related to a claim and evidence that is not
- Distinguish between opinion and evidence in an argument

For example, the teacher may design an investigation in primary school (grades K–2) that involves solving the design problem "Find a design problem in a picture book and make a claim about the problem in the book using the evidence in the book to support your definition of the problem. Use materials to design a solution to the problem by mimicking how plants and/or animals use their parts to help them survive and meet their needs. For example, plants use water to stay upright, tree nuts have hard shells to protect the inside, and lily pad leaves are flat and wide so they can float on water." The teacher can model how students can work collaboratively to find a design problem in a chosen picture book and make a claim to identify the problem, supported by evidence in the book. For example, students may choose the story "The Three Little Pigs" and the teacher can make a claim that the straw house needs some reinforcement so the wolf cannot blow it down. The teacher can list details about the pictures in the book and the story to support their claim, pointing out how to use supporting evidence to make a claim. Students can listen to the argument and identify the main points of the argument when the teacher models the argument. Teachers can ask students to explain their notes about the main points of an argument to check if they were paying attention to the key information.

10.3.2 Emulation

It is during this second step of the coaching strategy that the shift from teacher-led to student-led activities begins. The emulation step is related to the SRL phase of forethought (Figure 10.1), but is different from modeling because it guides students to be the initiator of their learning about engaging in argument using evidence. During emulation, the role of the student is to replicate the science and engineering practices that the teacher models when they are given a similar task as the model. However, the student does this with considerable support, and the teacher provides the students with a checklist to set goals, and consider their self-beliefs regarding a science and engineering practice in the investigation. The checklist should begin with the outcome goal and then list the process goals that will help students define the problem they are going to pursue in the investigation.

In the emulation phase, students can continue in the engineering design process to ideation of how they can mimic plants or animals to solve their design problem. For the three little pigs' straw house example, the teacher can show how a tree nut has a hard shell to protect the soft inside and use the same design to create a shell for the straw house. The teacher can present evidence for this design idea that is related to the plant mimicry and some evidence that is not related to the plant mimicry. The teacher can then guide student thinking to sort through the evidence related to the claim and the evidence that is not related to the claim.

Students can then work in small groups to make claims about their own chosen books about what is a design problem, supported by details in the book. When students have finished with their work, they can present their arguments to the class and the class can agree or disagree with the presentations, supporting their positions with evidence. If students make a claim that is not supported by evidence, the teacher can guide their thinking to identify arguments that are supported by evidence.

Students can use the following checklist that supports some of the learning tasks. Note that this list is shorter because it is for younger students.

- I understand that the goal is to design a solution for a problem in my book (Outcome goal)
- I found a problem that would solve something in my book
- I can explain the problem I found to someone else
- I can use details in the book to show that it is a problem
- I can build something that copies the way a plant works from the problem in the book

Students use these statements to help them improve their self-efficacy for engaging in arguments using evidence. Like cognitive apprenticeships, MPI-S helps students who may not have had prior access to ways of knowing in science and engineering by explicitly pointing out not just how to do the practice (procedural knowledge), but also why they are doing the practice (epistemic knowledge). In later lessons, teachers can use different checklists for different science and engineering practices.

10.4 Assessing Students in Self-Regulating Their Learning about Engaging in Arguments from Evidence

10.4.1 Self-Control

The self-control step of MPI-S is related to the performance phase of SRL (Figure 10.1) because it helps students monitor their performance and focus their attention on learning about science and engineering practices. Students engaged in MPI-S to this point have observed what they are supposed to be accomplishing through the teacher model (Modeling) and have attempted similar skills and knowledge with support from the teacher (Emulation). In the third step, the teacher continues to support student

Engaging in Argument from Evidence

self-regulation of learning about science and engineering practices but reduces support to allow students to actively reflect on their metacognitive strategies. The third and fourth steps of this coaching approach can also be used to assess how well students are beginning to self-regulate their new skills. Teachers should provide students with a more difficult attempt at the skill they are trying to build and give students only a few basic standards from which to check.

When students identify their design solution, they can make a prototype and test it to see if it solves the problem. Students can gather data and present the interpreted data as an argument to the class for peer review. During this peer review, students can distinguish between an opinion and evidence in their argument.

For the science and engineering practice of engaging in arguments using evidence, teachers support students by providing the following shortened checklist during their peer review to stay focused on making a claim with relevant evidence rather than opinion.

- I understand that the goal is to show how our data from our prototype solves the problem in our book (Outcome goal)
- I can explain the evidence we gathered to another person
- I can use the evidence to make a claim about how our design solution works

To check for appropriate metacognition in this step, teachers should ask a few questions about the choices that students make when they perform the practice such as:

- How do you know your design solution solves the problem in the book?
- How do you know you have evidence and not opinion to support your claim?

If students struggle to answer these questions, then they likely have not mastered the learning tasks of the practice. Teachers can help students by going back and modeling the difficult practices, then giving students another chance to emulate. When students can answer these questions in a way that demonstrates their mastery of engaging in arguments using evidence, then teachers can fade all support in the next step, self-reflection.

10.4.2 Self-Reflection

In the self-reflection step of MPI-S, students perform the targeted practice entirely on their own and reflect on the outcome. This last step of MPI-S is aligned to the self-reflection processes of SRL (Figure 10.1) and helps students to self-evaluate their performance and attribute their successes and failures to sources of learning. The self-reflection step builds upon the self-control step because students are expected to regulate their learning without any support. Students should be able to demonstrate they can both understand and implement the science and engineering practice (engaging in argument using evidence) without any teacher support. In this step, the teacher gives the learning task and ensures that the student was able to accomplish it in a way that parallels the science and engineering practice. The teacher can decide if the student has mastered the science and engineering practice by evaluating student answers to the questions about their rationale. Questions that a teacher would ask regarding engaging in arguments using evidence for primary school students engaged in design solution for a problem in a picture book investigation are:

- How did you know if something is evidence or opinion? (Goal setting and strategic planning)
- Explain how you noticed if a design solution was going to solve a problem. (Metacognitive monitoring)
- Explain how you used your data to make a claim about how well your solution solved the problem. (Attention focusing)
- What about making claims with evidence do you think you did well? Why? (Attribution)
- What about making claims with evidence do you think you still need to practice? Why? (Attribution)
- What would you change about the way you make claims with evidence in the future? Why? (Adaptation)

In the next section, there will be two case studies to demonstrate how different processes of self-regulated learning could help or hinder students in their learning while they engage in arguments using evidence.

10.5 Teacher–Student Cases

10.5.1 Case Study: Adaptive

The case study featured in this chapter as an example of an adaptive and effective learner for engaging with arguments using evidence will focus on:

- Self-efficacy in the forethought phase
- Metacognitive monitoring in the performance phase
- Adaptation in the self-reflection phase

Nina, a 1st grade student, is working with her group to define a problem in the picture book her group chose from the teacher's list. Nina knew that she needed to find details from the book to explain the problem (making a claim with evidence), so she felt confident because she knew what she was supposed to do. Nina was going to ask the teacher if she was right in understanding what she was supposed to do, but wanted to try to do this on her own. Instead, she checked her understanding of the outcome with her group and her group agreed with her understanding, which helped improve her self-efficacy of the learning task. Nina approached her learning task with the idea that most stories had an interesting part where a character had to change something. She planned to make a list of changes that needed to be made in the story and from there she was going to find a problem that they could solve by copying the way a plant survives.

When Nina and her group were going through the story to define a design problem, they made a list by drawing out ideas from the study when something needed to be changed. They found three times something needed to be changed in their book. Nina looked at her checklist and thought about how she can do each bullet point on the checklist. She knew that she could find a problem that would solve something in her book because her group drew out the three problems (metacognitive monitoring). She then asked her group to take turns explaining the three problems so that she could do the second bullet point on the checklist. The group explained the problems, and at the end of the activity realized that they need to pick one of the problems to focus on their design solution. Nina suggested that the solution should be something that satisfies two conditions: that it was buildable and that it needed to copy a plant. She knew these things because she knew the ideas on the checklist were there to help her get to a solution (metacognitive monitoring). Nina's group reviewed the three problems they identified in the story and decided that only one of them could be answered by building something that had the same characteristics as a plant. The other two were character conflicts that needed to be resolved with communication. Her group was able to make a claim about their design problem using the details they drew out from the book.

After the students presented their ideas using details from the book, they received the feedback that they missed one detail from the book that could have helped them define their problem and that two of the ideas were not details from the book but were their own opinions from prior experiences. Some of the other students in the group had hurt feelings by this feedback, but Nina thought they were right. She decided that they had a point because she knew what to focus on and realized that their evidence was not entirely from the story.

She also had enough self-efficacy for the task that she was not personally hurt by her peers' and her teacher's feedback. Nina thought about what more she could do than just follow the checklist. She decided that she was going to draw or write what the checklist asked her to do right under the bullet point, so then she could line up her ideas with the help the teacher was giving her.

10.5.2 Case Study: Maladaptive

The case study featured in this chapter as an example of an ineffective learner that needs support for asking questions will focus on:

- Self-efficacy in the forethought phase
- Metacognitive monitoring in the performance phase
- Adaptation in the self-reflection phase

Gabby, a 1st grade student in Nina's class, is also working with her group to define a problem in the picture book her group chose from the teacher's list. Gabby was very shy and was still having a hard time talking to more than one person at a time. Gabby really did not understand what they were supposed to do when the teacher said to find a problem in the book, so Gabby looked to see if any of the characters were upset because that usually went along with a problem. As a result, Gabby thought that her task was to make a list of the characters who were upset in the story.

When Gabby started checking what she did to make her list with her peers (metacognitive monitoring) she asked one person what they did. She did notice that her list of problems did not look like lists from the other students in the group, but she was too shy to speak up. Her lack of self-efficacy for the task halted her ability to do anything about her metacognitive monitoring. Gabby decided that she had a list to turn in and that was good enough.

Gabby's group also received feedback that they had used opinions rather than evidence to make a claim about the problem they identified in the story. Since Gabby had an entirely different idea about what the outcome should be for this project, she just shrugged off the feedback and thought she might pay attention the next time. She felt this time that she was too far off course to change anything and wanted to just forget the whole experience.

10.5.3 Supports the Teacher Can Provide for Gabby in This Case Study

10.5.3.1 Self-Efficacy

The teacher knew Gabby was shy and was only comfortable talking with one person at a time. The teacher could have talked to Gabby about her self-efficacy one-on-one by asking her about her prior experiences with picture books and how she interpreted them. This could have given the teacher some information about how Gabby sees claims and evidence as well as her reading comprehension. The teacher could have done some extra modeling for Gabby based on Gabby's responses to her question. This could have had an effect of raising Gabby's self-efficacy as long as the teacher could show Gabby that she has the skills to identify a problem in the book and use details to explain why it was a problem.

10.5.3.2 Metacognitive Monitoring

If Gabby had started with higher self-efficacy, it may have improved her ability to act on her metacognitive monitoring. Gabby decided not to change course even though she did the first half of metacognitive monitoring (recognizing her work was not meeting the goal) because her self-efficacy was low. With a few tweaks, Gabby may have been confident enough to make a few tweaks to her work when she realized she had a different understanding of the process than the other students in her group.

10.5.3.3 Adaptivity

Having an environment where it is safe to make mistakes might have helped Gabby be more adaptive when she received feedback that her work did not meet all of the process goals or the outcome goal. However, having a classroom where students feel safe to make mistakes might not have influenced Gabby because her self-efficacy was so low and she was reluctant to talk to many people. The teacher again could have had a one-on-one conversation with her or had a trusted peer discuss how they make changes when they realize they made a mistake. If Gabby mentioned her ability to see that she was on the wrong path during her metacognitive monitoring in the performance phase, the teacher could have used that opening to explain that making mistakes is a part of learning and that making changes in your learning processes is important when you are learning something new. Gabby might see that it is not embarrassing to change the way you learn when it is not working well.

Reflection question for teachers: Reflect on the adaptive and maladaptive case study you just read and answer the following question. What student supports for asking questions can you use in your classroom from these case studies?

10.6 Designing Lessons That Use the Practice of Engaging in Arguments from Evidence

In order to plan for teacher modeling of the practice "engaging in arguments using evidence," teachers can review a lesson plan for ways they can model the key learning tasks for that practice in an investigation. Teachers can look for opportunities in their lesson to point out to students:

- How to make a claim using evidence
- Ways to see if evidence is relevant or not to a claim
- Ways to distinguish between evidence and opinion
- How to listen to an argument and point out the main features of an argument

Teachers can then make checklists for students by listing the key features of doing the science and engineering practice, and then convert them into student-friendly language for a checklist bullet. For example, a key feature from engaging in argument using evidence is to listen to an argument and point out the main features of the argument. The checklist bullet item for this key learning task can be "I can make a model drawing of the parts of an argument and identify the claim and evidence." Teachers should make at least one bulleted item for each key feature of the practice. Making more than one item could potentially help students by understanding the practice from different perspectives.

10.7 Teacher Reflection on Implementation of Lesson Featuring Engaging in Argument from Evidence

Being a reflective practitioner is important for supporting student self-regulated learning. Teachers should consider the following questions for the practice of engaging in argument using evidence:

- What worked well in supporting students' engagement in arguments using evidence?
- What did not work well in supporting students' engagement in arguments using evidence?
- How will I change this the next time I teach students to engage in arguments using evidence?

CHAPTER 11

Evaluating and Communicating Information

Each of the chapters in Chapters 4 through 11 has a parallel design that discusses a particular science and engineering practice in detail. This chapter focuses on teacher support for students evaluating and communicating information in science and engineering. In each chapter, the practice is dissected into distinct and clear learning tasks. These tasks are then examined within the context of a self-regulated learning cycle. A multistep coaching strategy is explained and points for instruction and assessment are given using the example of a design challenge for students in grades 3 through 5 to improve the school recycling program. The tasks are reassembled into two case studies – one positive and one negative – to demonstrate how the learning tasks can be used by students.

11.1 Learning Tasks in the Practice

The other practices listed in the Next Generation Science Standards (NGSS Lead States, 2013) tend to be somewhat sequential in an investigation, although there are always some iterative and overlapping processes among the practices. The practice of evaluating and communicating information is woven throughout the other practices. This practice can be accomplished without doing an investigation, because one may be communicating about or evaluating an investigation done by other people. A scientist or engineer may ask questions or design solutions, then create a conceptual model, then proceed to creating and implementing a procedure to test the model, collecting data using mathematical and computational thinking, then analyze and interpret the data, and either explain the findings or create a claim using evidence. Evaluating and communicating science is a skill that is associated with all of the prior practices, but it can also stand alone. For example, a person may try to find out how to build a shed and need to read or watch videos on the design and construction, evaluating reliable sources, and communicating the plan and process to a colleague.

The Next Generation Science Standards (NGSS Lead States, 2013) also states that the practices of obtaining, evaluating, and communicating information are done by both scientists and engineers. First, ideas and findings must be communicated precisely and clearly so that others can judge the trustworthiness of the information. Second, the science and engineering profession has expectations that people will work collaboratively to answer questions and solve problems, thus the need for effective and valid communication of information is necessary for collaboration (Stehle & Peters-Burton, 2019). Third, the use of multiple sources of information is a foundation of science and engineering disciplines, so skills involved in obtaining, evaluating, and communicating are important for flexible thinking and making connections.

These overall definitions have been further broken down into grade bands by NGSS of K–2 (aged 4 to 7), 3–5 (aged 8 to 11), 6–8 (aged 12 to 14), and 9–12 (aged 15 to 18). The following sections explain smaller learning tasks for each grade band that comprise the science and engineering practice of evaluating and communicating information. The sections begin with the learning tasks from K–2 and build on the new practices that can be learned in the successive grade band (NGSS@NSTA, 2014). From there, the learning tasks are decomposed, which represent what students should be able to do to master the practice and can also be used as process goals for students. The chapter also provides examples of how a teacher can support the learning task in an explicit way and in a reflective way. Educational research on epistemic knowledge in science has shown that students are less likely to learn an aspect of the nature of science implicitly, but are more likely to learn that aspect when taught explicitly by the teacher and when given intentional opportunities to reflect on how the aspect was demonstrated (Khishfe & Abd-El-Khalick, 2002; Peters, 2012; Peters & Kitsantas, 2010; Peters-Burton, 2015). This literature forms the basis for ideas to offer practical ways to enhance student learning for science and engineering practices.

Breaking the practice down is not only important for helping students who may not have had exposure to doing science and engineering, but it is also important to situate the learning task in a cycle of self-regulated learning. In order for self-regulated learning to be productive, one must begin with a well-defined learning task. As explained in Chapter 3, goal setting is key to being successful with self-regulated learning, and the goal needs to be discrete and composed of smaller, more proximal process goals that lead up to the outcome goal.

11.1.1 Process Goals for Evaluating and Communicating Information

Educational research has demonstrated that students can more effectively learn when teachers explicitly and reflectively support learning tasks (Peters-Burton, 2017). Teachers can model the behaviors that are shown by scientists and engineers when performing the practice of evaluating and communicating information in the course of teaching the practice explicitly. When reflectively teaching the practice, teachers give students a chance to show what they know about performing the practice. From assessing the student's reflective response, teachers can determine if they need to reteach the practice or give students another chance to emulate the practice.

11.1.1.1 Explicitly and Reflectively Supporting Process Goals for Evaluating and Communicating Information for Grades K–2

The Next Generation Science Standards explain that this practice of obtaining, evaluating, and communicating information for students in K–2 builds on prior experiences and uses observations and texts to communicate new information (NGSS@NSTA, 2014). The following learning tasks can be used as process goals to instruct students about the detailed skills needed to master this practice:

- Use media or age-appropriate text to identify scientific information that can be used as evidence for a scientific phenomenon
- Use media or age-appropriate text to identify technical information that can be used as evidence for a design solution
- Describe how an image supports a design solution or a scientific phenomenon
- Use text features (headings, figures, tables) to identify scientific or technical information
- Communicate information (written or orally) that provides details for a scientific phenomenon or design solution

Teachers can explicitly teach students how to *use media or age-appropriate text to identify scientific information that can be used as evidence for a scientific phenomenon*. In full view of the whole class, demonstrate to students the expectations for identifying scientific information by talking aloud while reading the text or engaging with the media. Pause while reading or engaging and discuss aloud pieces of information at these intervals so that the whole reading is broken into smaller parts. To help students reflect on their progress in learning the skill, give students an age-appropriate text or media piece that contains scientific information. Ask students to highlight the scientific information by writing or making a drawing and explaining where it was found in the text or media piece.

Teachers can explicitly teach students how to *use media or age-appropriate text to identify technical information that can be used as evidence for a design solution* so that students can understand the processes involved in the skill. In full view of the whole class, demonstrate to students the expectations for identifying technical information by talking aloud while reading the text or engaging with the media. Pause while reading or engaging and make deliberate decisions about pieces of information along the way. To assist students in reflecting on their skills, give students an age-appropriate text or media piece that contains technical information. Ask students to highlight the technical information by writing or making a drawing and explaining where it was found in the text or media piece.

By explicitly teaching students how to *describe how an image supports a design solution or a scientific phenomenon*, teachers can model disciplinary behaviors. Present an image to the class and model how you would look at the image through perspectives and deliberately gather the information from the image. Teachers can help students reflect on their progress by giving them an image to interpret for scientific or technical information. Ask students to explain how they will interpret the information to a peer and have a discussion of their different perspectives.

Teachers can model for students how to *use text features (headings, figures, tables) to identify scientific or technical information*. Present a text selection to the class and model how you would look at the text to identify headings, figures, tables, and other features. Explain your thinking about which information is scientific or technical and which is not. Then help students to reflect on their skills by giving them a text to interpret for scientific or technical information. Ask students to explain how they will interpret the information to a peer and have a discussion of their different perspectives.

Teachers can explicitly model ways through which students can *communicate information (written or orally) that provides detail for a scientific phenomenon or design solution*. Select a conceptual model that has been used before for an investigation. Use this model to communicate the scientific or technical information to students, deliberately making mistakes. When making mistakes, explain to students how you recognized that you made a mistake and how you will correct it. Then, have students perform the skill and reflect on their progress. Put students in pairs or small groups and give each of them a different conceptual model to explain. Ask students to take turns explaining their model to other group members. Have the group members ask clarification questions after the explanation. When the discussion is over, ask students to write a reflection note on how they communicated the information.

11.1.1.2 Explicitly and Reflectively Supporting Process Goals for Evaluating and Communicating Information for Grades 3–5
According to the Next Generation Science Standards, obtaining, evaluating, and communicating information for students in 3–5 builds on K–2 experiences and progresses to evaluating the merit and accuracy of ideas and methods. The following breakdown of learning tasks can be used as process goals to guide students to understand the skills that make up the practice:

- Summarize scientific or technical information from age-appropriate texts or media
- Summarize scientific or technical information across multiple age-appropriate texts or media
- Use text features (headings, figures, tables) across multiple sources to summarize scientific or technical information
- Obtain reliable technical or scientific information from multiple sources
- Put forth an argument for why a particular resource is reliable for technical or scientific information
- Communicate scientific or technical information to other people using charts, graphs, and figures

Teachers can explicitly teach students to *summarize scientific or technical information from age-appropriate texts or media* so that they can see the process in action. Present text or media to the class and methodically go through the material, noting the details that are scientific or technical. From these details model your thinking for students on how to create a summary of the ideas. Then, teachers can help students reflect on their performance of the skill by giving them text or a media selection and asking them to summarize the scientific or technical information in the text or media. Ask students to explain how they know they captured all of the information that is scientific or technical from the selection.

Teachers can model how to *summarize scientific or technical information across multiple age-appropriate texts or media* for students. Present multiple sources of text or media selections to students and methodically go through the material, noting the details that are scientific or technical.

From these details model your thinking for students on how to create a summary of the ideas. Then, help students to reflect on their learning of the skill by giving them multiple sources of text or media selections and asking them to summarize the scientific or technical information in the text or media. Ask students to explain how they know they captured all of the information that is scientific or technical from the selections.

Teachers can model for students how to *use text features (headings, figures, tables) across multiple sources to summarize scientific or technical information*. Display a text with different features to the whole class. Demonstrate to students how you would systematically go through headings, images, tables, and figures to gather information from the text. Perform the same modeling with other texts and demonstrate to students how you would summarize the ideas gathered from the texts. Then ask students to reflect on their progress in learning the skill by giving them multiple texts from which to use features. Show students a way to make notes about what information they are gathering from the text features, such as a graphic organizer. Ask students to summarize their information.

Students can learn strategies for *obtaining reliable technical or scientific information from multiple sources* when a teacher models the skill. Search for technical or scientific information from multiple sources, noting how you found the sources and how you are evaluating the reliability of the source by explaining your thinking about the expertise, relevance, timeliness, accuracy, and purpose of the source. Then help students to reflect on their ability by asking them to search for scientific or technical sources and explain the expertise, relevance, timeliness, accuracy, and purpose of each of the sources.

Teachers can explicitly teach how to *put forth an argument for why a particular resource is reliable for technical or scientific information*. Using the criteria of expertise, relevance, timeliness, accuracy, and purpose of the source from the prior learning task, demonstrate to students how you make a claim, back the claim with evidence, and use reasoning of the criteria standards to support your claim. After a student practices the skill, teachers can set up a learning environment for students to reflect on their progress. From the resources that students wrote about in their graphic organizer from the search (task from above), ask students to write a claim about the reliability of their sources, backing it up with evidence and reasoning.

Teachers can help students understand what is involved in *communicating scientific or technical information to other people using charts, graphs, and figures* by modeling it. Gather a selection of charts, graphs, and figures.

Demonstrate to students how you would communicate scientific or technical information to them by talking aloud your thoughts. When students get a chance to practice the skill on their own, teachers can help students be reflective about their learning. Put students in small groups and give them a selection of charts, graphs, and figures. Ask students to take turns for communicating the scientific or technical information to the other students. Explain that they should communicate as much information from the charts, graphs, and figures as accurately as they can.

11.1.1.3 Explicitly and Reflectively Supporting Process Goals for Evaluating and Communicating Information for Grades 6–8

According to the Next Generation Science Standards, obtaining, evaluating, and communicating information for students in grades 6–8 builds on K–5 experiences and progresses to evaluating the merit and validity of ideas and methods. The following learning tasks can be used as process goals for students to understand the underpinnings of the practice:

- Identify scientific or technical ideas in an age-appropriate technical text
- Evaluate an age-appropriate text that contains scientific or technical information for credibility, accuracy, and bias
- Integrate qualitative and quantitative information to explain a scientific phenomenon or design solution
- Evaluate scientific or technical information in light of contrasting information
- Communicate scientific or technical information from a diagram in writing or orally

Teachers can explicitly teach how to *identify scientific or technical ideas in an age-appropriate technical text* so that students can see how the process is accomplished. In full view of the whole class, demonstrate to students the expectations for identifying scientific information by talking aloud while reading the technical text. Pause while reading or engaging and make deliberate decisions about pieces of information along the way. To help students in their reflection, give students an age-appropriate technical text that contains scientific information. Ask students to focus on the scientific information by writing or highlighting where it was found in the technical text.

Teachers can model for students how to *evaluate an age-appropriate text that contains scientific or technical information for credibility, accuracy, and bias*. Display a text to the whole class and read through it together. Make a chart to list the attributes of credibility, accuracy, and bias for the students.

Go through the article and fill in the chart for each, discussing it with the students. Create a rubric that students can use from the chart to guide their evaluation of texts for credibility, accuracy, and bias. Teachers can help students reflect on their learning of the skill by giving them a text and the rubric for evaluation of the text and ask them to fill out the rubric. Ask students to summarize their evaluation of the credibility, accuracy, and bias of the text.

Modeling how to *integrate qualitative and quantitative information to explain a scientific phenomenon or design solution* can help students see the strategies needed to perform the skill. Show students qualitative and quantitative information on a design solution. While you talk aloud about your thinking and organization, explain how you would go about integrating the information to show how the design solution works. Then to help students reflect on their learning, give them qualitative and quantitative information on a design solution and ask them to explain how the design solution works by integrating the information. Ask students to describe the process they used to integrate the information.

Teachers can explicitly show students how to *evaluate scientific or technical information in light of contrasting information*. Present two different arguments for a scientific phenomenon to students. Model for them how you would go about evaluating the scientific and technical information to make a decision on the trustworthiness of the information. Then after students get a chance to practice the skill, give them two contrasting pieces of information about the same phenomenon. Ask students to form an argument using a claim, evidence, and reasoning to explain which information is more credible and trustworthy.

Teachers can demonstrate their strategies for *communicating scientific or technical information from a diagram in writing or orally*. Display a diagram to the whole class and explain how you would go about identifying information from the diagram to communicate and how you could clearly communicate it, talking aloud about your thinking. Then to help students to reflect, place them in small groups and give them a diagram to describe in writing. When they finish, have the group peer review their writing about the diagram and have a discussion to reach a consensus about the most effective way to communicate information from the diagram.

11.1.1.4 Explicitly and Reflectively Supporting Process Goals for Evaluating and Communicating Information for Grades 9–12

The Next Generation Science Standards state that obtaining, evaluating, and communicating information for students in grades 9–12 builds on

Evaluating and Communicating Information

K–8 experiences and progresses to evaluating the validity and reliability of the claims, methods, and designs. The following learning tasks can serve as process goals for students so that they can learn the skills involved in the practice:

- Accurately paraphrase scientific or technical information from an age-appropriate text or media
- Address a scientific question or design problem using multiple sources of media and text, with a mind toward valid and reliable sources
- Evaluate the evidence given in an age-appropriate scientific or technical text
- Use multiple sources of valid and reliable information to make a claim that synthesizes the information
- Communicate scientific or technical information from multiple forms of resources using multiple types of communication

Teachers can explicitly teach students to *accurately paraphrase scientific or technical information from an age-appropriate text or media* by having students read a selected text. Once everyone has read the text, demonstrate to the whole class how you would go about paraphrasing information, making notes about the key processes in a systematic way. After students get a chance to apply the skill, then teachers can help them reflect. Place students in small groups and give them a text to paraphrase. When they finish, have the group peer review their writing and have a discussion about how they went about converting the text to the paraphrased writing.

Teachers can model how to *address a scientific question or design problem using multiple sources of media and text, with a mind toward valid and reliable sources* for students. Present students with a scientific question and display the way you search for texts and media to help answer the scientific question. Make a chart to note the results of each source with regard to relevance, timeliness, expertise, and accuracy. Talk aloud about your decisions as you go from the source to the notes in the chart. Demonstrate how you disregard the less credible sources and answer the question with the more credible sources. Then, after students have had a chance to perform the skill, teachers can help students reflect on their progress. Provide students with an empty chart with the categories of relevance, timeliness, expertise, and accuracy. Ask students to address a scientific question of your choice by finding information about the phenomenon. Ask students to fill out the chart to evaluate the articles while gathering information from the articles to answer the question. Ask students to write a summary about what sources they used and why and how they addressed the question with the information from the credible sources.

Teachers can model how to *evaluate the evidence given in an age-appropriate scientific or technical text* so that students understand what is involved in the skill. Select two texts for discussion, one that has credible evidence and one that does not. Demonstrate to students how to go about evaluating the evidence in the text, making notes about the key characteristics. Create a rubric using the key characteristics for evaluation with the class. When students have performed the skill in several settings, teachers can help them reflect on their learning. Provide a text to students along with the rubric that was developed with the class discussion. Ask students to evaluate the evidence in the text using the rubric, but also to write a reflection of their interpretation. Have students form small groups or a Socratic circle to discuss their interpretations and rubrics.

Modeling the *use of multiple sources of valid and reliable information to make a claim that synthesizes the information* can help students see how to perform the skill at disciplinary standards. Demonstrate to students how to extract information from multiple sources, such as journal articles, to make a claim that synthesizes the information. Demonstrate to students the evidence and reasoning you use to back up your claim. Teachers can help students reflect on their progress of learning the skill by providing them with multiple sources of valid and reliable information and asking them to make a claim that synthesizes the information in an essay that demonstrates the evidence and reasoning they are using to make the claim.

Teachers can model for students how to *communicate scientific or technical information from multiple forms of resources using multiple types of communication*. With a whole class, make a list of different ways to communicate scientific and technical information. Communicate the summary of multiple forms of information in as many formats as possible. Make a chart with the class to evaluate the benefits and drawbacks of each type of communication. Teachers can help students reflect on their learning by providing students with multiple forms of text and media information regarding a design solution. Ask students to use at least three different formats of communication to explain the design solution.

11.2 Strategies for Teachers to Support Student Practices

For the practices of "evaluating and communicating information," this section will focus on the SRL processes of *goal orientation* in the Forethought phase, *attention focusing* in the Performance phase, and *self-reaction and self-evaluation*

Evaluating and Communicating Information

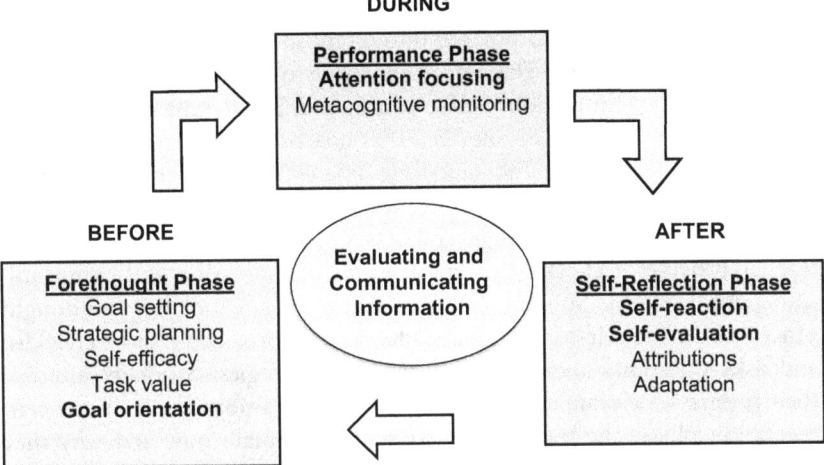

Figure 11.1 SRL processes for evaluating and communicating information

in the Self-Reflection phase as seen in Figure 11.1. The focus will be on these processes because they are most illustrative for the practices of evaluating and communicating information. In other chapters discussing other practices, the book will focus on other SRL processes so that the whole cycle will be discussed across the book.

11.3 Instructing Students in Self-Regulating Their Learning about Evaluating and Communicating Information

The Metacognitive Promoting Intervention-Science (MPI-S) is based on a research-based teaching strategy that supports student SRL development (Peters, 2009; Peters & Kitsantas, 2010). MPI-S enacts SRL because it prompts students about setting advantageous goals for learning a task by explicitly demonstrating the practice, which gives students who do not have experience with the practice an entry point to learning. MPI-S also assists students in focusing their attention on the key features of evaluating and communicating information, and asks students to reflect on their successes and failures in their performance of the practice and adapt accordingly. This teaching strategy works well for lessons involving science and engineering practices because it engages students with the processes and approaches to thinking rationally and systematically.

MPI-S is a suite of curricular tools, made up of a suite of checklists and questions that can be incorporated into established lesson plans to support student SRL strategies. The implementation of MPI-S consists of four steps: Modeling, Emulation, Self-Control, and Self-Regulation. The steps of MPI-S are the same ones as the coaching strategy founded by Zimmerman (2000). In the modeling phase, the teacher demonstrates key features of evaluating and communicating information. Students consider their own forethought processes. In the emulation phase, the teacher provides a checklist of key features of evaluating and communicating information. Students use the checklist as a tool for forethought processes. In the self-control phase, the teacher provides a short checklist and asks questions about student learning strategies. Students monitor their progress for evaluating and communicating information. In the self-regulation phase, the teacher asks students to explain how and why they used key features of evaluating and communicating information. Students identify the instances of learning and assess their quality of evaluating and communicating information.

Using this approach, the teacher initially supports students explicitly through modeling and then drops the level of support so that students are able to articulate how they understand how to evaluate and communicate information (or any of the science and engineering practices) independently. The first two steps of MPI-S (modeling and emulation) are instructional and the second two steps (self-control and self-regulation) assess student learning to inform instruction. MPI-S does not increase the time it takes to teach science and engineering practices, because the teaching focuses on student use of science and engineering practices that are tangible, and at the same time they are creating opportunities for students to be more aware of their learning strategies.

11.3.1 Modeling

Students averse to risk in their learning may take on a performance goal orientation, which emphasizes getting through the learning activity without appearing to fail or struggle (Dweck & Leggett, 1988). Students who adopt a mastery approach to learning are not afraid to fail or demonstrate publicly that they are struggling because their focus is to learn the task deeply.

The modeling step in MPI-S helps students with their forethought processes such as goal orientation because it demonstrates positive learning goals, and helps them evaluate their self-efficacy and value in the task. In this case, students may see that when the teacher models when they struggle or find points of difficulty in learning yet persist by finding another learning approach, it is not detrimental to fail or struggle. Struggle in learning can produce even more advanced learning outcomes because the learner needs to persist and find other ways to achieve. Therefore, teachers should take care to explain other applications of the skill of evaluating and communicating information so that students can have goal orientation beliefs about mastering the learning task. Teachers can also add other examples of valuable uses of the practice for students during later phases of the coaching strategy.

Modeling is much like a cognitive apprenticeship (Collins, Brown, & Newman, 1989) where the mentor (teacher) does the activities in full view of the apprentice (students), but at the same time talks aloud about rationale, choices, and decision points with the intention that the apprentice will be able to adopt the same practices. The role of the student in this first step of MPI-S is to notice key features of the skill as demonstrated by the mentor and ascertain the overall sense of the outcome. Students will learn how to reach the outcome through understanding process goals in later steps of the teaching strategy.

Recall from the Next Generation Science Standards that some of the key features of evaluating and communicating information that students in grades 3–5 are expected to master are:

- Summarize scientific or technical information from age-appropriate texts or media
- Summarize scientific or technical information across multiple age-appropriate texts or media
- Use text features (headings, figures, tables) across multiple sources to summarize scientific or technical information
- Obtain reliable technical or scientific information from multiple sources
- Put forth an argument for why a particular resource is reliable for technical or scientific information
- Communicate scientific or technical information to other people using charts, graphs, and figures

For example, the teacher may design an investigation in grades 3–5 that involves solving the design problem, *Find ways to expand the school's current recycling program while increasing costs by a maximum of 20 percent. To begin, you must first find out what the school is currently recycling, how they are doing it, and how much it costs. Then investigate materials and processes that the school is doing that could result in more recycling.* In order to accomplish this, students must engage in obtaining, evaluating, and communicating science and engineering information, specifically engaging with the skills mentioned above. The teacher can find information on local recycling programs and share some of the information with the students, such as types of plastic that can be recycled locally and how the plastic must be prepared. The teacher can first demonstrate how to evaluate the reliability of the information on plastic recycling by finding the source of the information and determining how experienced the authors may be on the topic. The teacher can then also demonstrate how to cross-reference information with other known reliable sources to determine the accuracy of the statements about plastic recycling. Once the class is convinced that the source of information is trustworthy, the teacher can then demonstrate how to read the diagrams and graphics in the information before moving on to reading the narrative with the students. Finally, the teacher can take notes about key information by talking aloud about their decisions on central information (which is recorded) and extraneous information (which is not recorded).

11.3.2 *Emulation*

It is during this second step of the coaching strategy that the shift from teacher-led to student-led activities begins. The emulation step is related to the SRL phase of forethought (Figure 11.1), but is different from modeling because it guides students to be the initiator of their learning about evaluating and communicating information. During emulation, the role of the student is to replicate the science and engineering practices that the teacher models when they are given a similar task as the model. However, the student does this with considerable support, and the teacher provides the students with a checklist to set goals, consider their self-beliefs regarding a science and engineering practice in the investigation. The checklist should begin with the outcome goal and then list the process goals that will help students define the problem they are going to pursue in the investigation.

In the emulation phase, students can continue to find, evaluate, and summarize information they find on how plastic, metals, and other materials are recycled. Students can work in small groups to learn about approaches recycling centers employ to process materials to be used again. A checklist of the outcome goal and key process goals that will help them achieve the outcome goal can support student work, along with teacher

Evaluating and Communicating Information 199

guidance. If the school currently recycles only paper, the students can look into other materials that can be recycled by local centers. It is more efficient if students can work backward to find out what can be recycled within their budget and then find out ways the school can add to recycling other materials. Both learning about how certain materials can be processed in a recycling center and processes the school can adopt to expand their recycling program require obtaining, evaluating, and communicating science and engineering information.

Students can use the following checklist that supports some of the learning tasks.

- I understand that the goal is to find new ways for the school to recycle more materials while staying in the budget (Outcome goal)
- I check to see if the author has experience in the topic
- I check other known materials to see if the author is correct
- I can summarize main ideas from more than one source of text and media
- I can read key information from tables and graphs
- I can use headings, figures, tables, and graphs to summarize information
- I can explain the information in tables, figures, and graphs to my classmates

Students use these statements to help them focus on mastery of the skills involved in obtaining, evaluating, and communicating information. Like cognitive apprenticeships, MPI-S helps students who may not have had prior access to ways of knowing in science and engineering by explicitly pointing out not just how to do the practice (procedural knowledge), but also why they are doing the practice (epistemic knowledge). In later lessons, teachers can use different checklists for different science and engineering practices.

11.4 Assessing Students in Self-Regulating Their Learning about Evaluating and Communicating Information

11.4.1 Self-Control

The self-control step of MPI-S is related to the performance phase of SRL (Figure 11.1) because it helps students monitor their performance and focus their attention on learning about science and engineering practices. Students engaged in MPI-S to this point have observed what they are supposed to be accomplishing through the teacher model (Modeling) and have attempted similar skills and knowledge with support from the teacher (Emulation). In the third step, the teacher continues to support student self-regulation of learning about science and engineering practices but reduces support to allow students to actively reflect on their metacognitive strategies. The third and fourth steps of this coaching approach can also be used to assess how well students are beginning to self-regulate their new skills. Teachers should provide students with a more difficult attempt at the skill they are trying to build and give students only a few basic standards from which to check.

When students have identified reliable sources of how materials are recycled locally, they can begin to interpret key text features and communicate key ideas from the information. The teacher can help students organize their information with some guidelines on how to write key information from tables, figures, and graphs as well as narrative information.

For the science and engineering practice of evaluating and communicating information, teachers support students by providing the following shortened checklist so that students stay focused on reading text features, summarizing key information, and communicating key information.

- I understand that the goal is to summarize the most important information from different sources (Outcome goal)
- I can explain the title, X- and Y-axis, and trends on a graph to another person
- I can explain the main idea of a figure and then explain in detail to another person

To check for appropriate metacognition in this step, teachers should ask a few questions about the choices that students make when they perform the practice such as:

- How do you know your source is reliable?
- How do you know you are communicating the most important information from a resource?
- How do you know you have communicated all of the important information from graphs and figures?

If students struggle to answer these questions, then they likely have not mastered the learning tasks of the practice. Teachers can help students by going back and modeling the difficult practices, then giving students another chance to emulate. When students can answer these questions in a way that demonstrates their mastery of defining problems, then teachers can fade all support in the next step, self-reflection.

11.4.2 Self-Reflection

In the self-reflection step of MPI-S, students perform the targeted practice entirely on their own and reflect on the outcome. This last step of MPI-S is aligned to the self-reflection processes of SRL (Figure 11.1) and helps students to self-evaluate their performance and attribute their successes and failures to sources of learning. The self-reflection step builds upon the self-control step because students are expected to regulate their learning without any support. Students should be able to demonstrate that they can both understand and implement the science and engineering practice (evaluating and communicating information) without any teacher support. In this step, the teacher gives the learning task and ensures that the student is able to accomplish it in a way that parallels the science and engineering practice. The teacher can decide if students have mastered the science and engineering practice by evaluating students' answers to the questions about their rationale. Questions that a teacher would ask regarding evaluating and communicating information for primary school students engaged in a design solution to find out what the school does to recycle and ways they can expand the program while staying within a 20 percent budget increase are:

- How did you know if the information is reliable or not? (Goal setting and strategic planning)
- Explain how you noticed if you found the key information from a text or media source. (Metacognitive monitoring)
- Explain how you used a graph or diagram to communicate information. (Attention focusing)
- What about evaluating and communicating information you think you did well? Why? (Attribution)
- What about evaluating and communicating information you think you still need to practice? Why? (Attribution)
- What would you change about the way you evaluate and communicate information in the future? Why? (Adaptation)

In the next section, there will be two case studies to demonstrate how different processes of self-regulated learning could help or hinder students in their learning while they evaluate and communicate science and engineering information.

11.5 Teacher–Student Cases

11.5.1 Case Study: Adaptive

The case study featured in this chapter as an example of an adaptive and effective learner for engaging with arguments using evidence will focus on:

- Goal orientation in the forethought phase
- Attention focusing in the performance phase
- Self-reaction and self-evaluation in the self-reflection phase

Jin, a 4th grade student, is working with her group to find reliable resources and summarize the information on local recycling programs. She knows from her experience in reading groups in school that the other students in her group are faster readers than she is, but when she watched her teacher model how to read key text features like graphs and diagrams, she realized that she did not need to be a fast reader to do what the teacher was asking. She had a mastery goal orientation because she decided that even though she might not be the fastest reader in the group, she could still contribute and was confident with making mistakes in front of her group. In her experience, when she made a mistake, it helped her to strengthen her skill because she remembered how to correct it.

Next, Jin and her group worked together to summarize the key information from the sources. Jin kept the checklist her teacher gave her and marked off the activity when she accomplished it (attention focusing). Sometimes she would get excited about her new ability to read a graph and give her group too much information in the summary. However, when her group explained that they only needed the most important information, she asked them to explain, and made a note of what they said on her checklist. She used her group's comments as additional items on her checklist and soon she was able to give just the right amount of detail when explaining a graph for a summary (attention focusing).

Finally, Jin is given some resources about her school's recycling program that explains what they have done over the past few years and how much it costs. She and her group have summarized the information and found some ways that the school could recycle water bottles from the cafeteria at the local recycling center and the only costs were to buy recycling bins. The teacher invited a community group to give students feedback. The community group gave Jin's group the feedback that they met the expectations of the project and that its expansion could be done easily. The community group also told them that since they had more funds to spend they could have looked for more opportunities beyond plastic bottles. Jin felt that this feedback was fair and trusted that the community group was trying to give feedback that would help them learn rather than making a personal judgment about them (self-reaction). Jin reflected on her learning in the project and felt confident that she could read a graph again and summarize information to explain only the key points (self-evaluation). She also understood that they had more funds in the budget and could have found another material to recycle, so she kept that in mind for the next time she does a project. She was sure that her design solution stayed within the constraints but felt like she might be more adventurous next time (adaptation).

11.5.2 Case Study: Maladaptive

The case study featured in this chapter as an example of an ineffective learner that needs support for asking questions will focus on:

- Goal orientation in the forethought phase
- Attention focusing in the performance phase
- Self-reaction and self-evaluation in the self-reflection phase

Sam, a 4th grade student in Jin's class, is also working with his group to find reliable resources and summarize the information on local recycling programs. However, Sam was thinking more about his upcoming soccer game that weekend than his school work. Sam figured he was going to be a professional soccer player and he would not need to know science and engineering to accomplish this goal. He did know that he had to pass all of his classes or his parents would not allow him to play soccer. Therefore, Sam was interested in getting a good grade on the project, but not as interested in internalizing and mastering the skills (performance goal orientation). Sam responded by listening to what his group was doing and he would copy their behavior and language, but did not think about why or how he was learning the skills of evaluating and communicating information.

Next, Sam and his group worked together to summarize the key information from the sources. Because Sam was focused on getting a good grade but not necessarily mastering the skills, he was mimicking his group's behaviors so he looked like he knew what was going on. He just did not feel that he needed to put in the effort really knowing about how to summarize the key information from the sources, so he let his group do much of the work and he copied what they did with one of the articles.

He was not focusing on the most important process goals in the skill, so his attention was on how to copy what his group mates were doing.

Finally, Sam is given some resources about his school's recycling program that explains what they have done over the past few years and how much it costs. Sam and his group found ways that the school could recycle cardboard, plastics #1 and #2, and aluminum cans in addition to the paper that was already being recycled. They had stayed within the 20 percent budget increase for the process as well. The community group gave Sam and his group feedback that they had accomplished all of the goals for the project and were very effective in their strategies for finding new ways to recycle at school. All Sam knew was that he received a good grade on the project and was not really interested in anything else (self-reaction). Because he met the goals he set for himself (get a good grade) he did not spend time and effort to reflect on his performance in the project (lack of self-evaluation). The next time he attempted a design challenge project, he did not have any additional strategies or tactics to help him with his learning.

11.5.3 Supports the Teacher Can Provide for Sam in This Case Study

11.5.3.1 Goal Orientation

The teacher knew that Sam was more interested in sports than his academic studies, so the teacher could have used Sam's interests to help him see that mastering the skills of evaluating and communicating information was important for everyone. The teacher could have helped Sam to be more motivated to master these skills by perhaps talking to Sam about the information he needs to possess to be a good soccer player and teammate, such as a rule book. The teacher could show Sam that in order to remember everything he needs to know, he needs to be able to pick out the most important information from the rule book, and he could learn to do that more effectively by fully participating in this project.

11.5.3.2 Attention Focusing

As a teacher, one always runs the risk of some students being carried by other students' efforts when working in groups. One way to guard against a group member riding on the coattails of other group members is to set up a design challenge to be interdependent across group members. The design challenge could have been adapted to equally divide up the reported summaries across group members. Additionally, group members could be held responsible for justifying the information in their summaries to be the most important information. In this way, the teacher could set up the assignment so that each student needed to accomplish the task successfully for the whole group to be successful.

11.5.3.3 Self-Reaction and Self-Evaluation

To help students become more aware of their learning processes, the teacher could have had a formal reflection component in the project. The teacher could have asked students to reflect on the feedback they received from the community group by asking them:

- How did you feel when you first received feedback from the community group? (Self-reaction)
- Summarize what you heard as feedback from the community group (Self-reaction)
- What did you do well to meet the goals of the project? (Self-evaluation)
- What do you want to change the next time you do a project like this one? (Self-evaluation and adaptation)

Asking these questions may have cued Sam that he should be thinking more deeply about his learning processes. If Sam answered them truthfully, then the teacher could have noticed that Sam is appearing to do well, but could use some practice in being more aware of his learning in the future.

Reflection question for teachers: Reflect on the adaptive and maladaptive case study you just read and answer the following question. What student supports for evaluating and communicating information can you use in your classroom from these case studies?

11.6 Designing Lessons That Use the Practice of Evaluating and Communicating Information

In order to plan for teacher modeling of the practice "evaluating and communicating information," teachers can review a lesson plan for ways they can model the key learning tasks for that practice in an investigation. Teachers can look for opportunities in their lesson to point out to students:

- How to read a graph, figure, or table
- Ways to evaluate the trustworthiness of an information source
- Ways to summarize key information from a single source
- Ways to summarize key information from multiple sources

Teachers can then make checklists for students by listing the key features of doing the science and engineering practice, and then converting them into student-friendly language for a checklist bullet. For example, a key feature from evaluating and communicating information is to identify key information from a graph. The checklist bullet item for this key learning task can be "When reading a graph, I read the title and the axes before I find a pattern in the graph. Then, when I find a pattern, I can explain the units as well as the pattern." Teachers should make at least one bulleted item for each key feature of the practice. Making more than one item could potentially help students by understanding the practice from different perspectives.

11.7 Teacher Reflection on Implementation of Lesson Featuring Evaluating and Communicating Information

Being a reflective practitioner is important for supporting student self-regulated learning. Teachers should consider the following questions for the practice of evaluating and communicating information:

- What worked well in supporting students' evaluation and communication of information?
- What did not work well in supporting students' evaluation and communication of information?
- How will I change this the next time I teach students to evaluate and communicate information?

PART III

Educational Research and Teacher Education Applications

CHAPTER 12

Professional Development Designs

This chapter examines the research associated with professional development settings for science and engineering practices and self-regulated learning. Since professional development tends to be developmental, the research reviews are separated into preservice preparation and inservice development. Each section of the chapter follows with a summary of recommendations derived from the research for preservice teacher instruction and for inservice teacher professional development experiences. For example, elementary teacher professional development for teaching data practices on the topic of earth sciences and secondary teacher professional development for teaching argumentation in science using SRL are described.

12.1 Preservice Teachers Instruction on Science and Engineering Practices

Because they are novices in their teaching careers, preservice teachers have not yet spent a great deal of time participating in and thinking about how students participate in science and engineering practices. They may have had their own learning experiences in science and engineering practices, but have not yet thought about how to translate them into teaching experiences for their future students. Although preservice teachers may not have extensive experiences, the following studies give information about what preservice teachers may be able to initially understand about science and engineering practices and what they may struggle with before moving on to becoming full-time teachers.

As discussed in earlier chapters, a learner who has high task value for a topic will tend to persist when faced with difficulties. Learners who have high self-efficacy for a particular learning task will tend to take on more challenges in their learning. Flores (2015) measured the personal self-efficacy and outcome expectations of thirty preservice elementary teachers before and after they participated in a fifteen-week science methods course where the first ten weeks consisted of learning new techniques for teaching practices and the last five weeks consisted of teaching fifth graders these practices. The elementary preservice teachers reported significant gains in personal self-efficacy for teaching practices after the methods course, but did not have significant gains in their outcome expectations. Hang and Srisawasdi (2021) explored preservice secondary teacher (n = 187) beliefs about the importance of the eight science and engineering practices in a survey. They found all of the preservice teachers rated all eight practices as important, with modeling having the lowest score within this range.

French and Burrows (2018) studied the work of thirty-eight preservice secondary teachers who underwent eighty hours of authentic science inquiry instruction covering two university courses. They examined the preservice teachers' interpretation of their ideal authentic science inquiry lessons before and after the two courses using a modified rubric analyzing science and engineering practices that were developed by Spuck (2014). They found that the teachers implemented the science and engineering practices of analyzing and interpreting data, conducting explanations, and communicating results the most. The categories of practices that the preservice teachers' lessons implemented the least were defining engineering problems and designing engineering solutions. The preservice teachers were proficient at writing inquiry-based lessons where they planned opportunities for their future students to collaborate, use scientific instrumentation, and collect and analyze data. However, they found that the preservice teachers needed additional support for developing student activities where students create testable questions, revise their question and methods, participate in peer review, and disseminate their results to their peers or the larger scientific community (French & Burrows, 2018).

12.2 Preservice Teachers Instruction on SRL

In the past, there has been some literature indicating it may take years for preservice teachers to integrate advanced techniques into their classrooms (Fuller & Bown, 1975). However, there is a growing body of literature that does not support this idea. This body of literature supports the preparation of teachers in more advanced ways of teaching before they enter the classroom, such as including the teaching of SRL for their students. Bråten and Strømsø (2005) examined the epistemological beliefs in motivational and strategic components of SRL with 108 student teachers. Using multiple regressions, they found that beliefs about knowledge construction were a strong predictor of SRL for the preservice teachers. This indicates that if preservice teachers are open to constructivist ways to teach, they may be ready to employ SRL in their future classrooms. Further, there may be a barrier that can be lifted that can help preservice teachers to learn more about teaching SRL before they take over their own classes: beliefs of the teacher educator. Tran et al. (2022) found that preservice teachers appreciated the role of SRL and tried to integrate SRL processes into their science lessons but needed further support from the teacher educator to fully implement SRL instruction more systematically. Kremer-Hayan and Tillema (1999) interviewed ninety teacher educators and preservice teachers and found differences in how these two groups viewed the meaning and implementation of SRL in the classroom. The teacher educators reported having a less positive attitude toward SRL and lower expectations about their competencies related to self-regulation than the preservice teachers. Perhaps preservice teacher educators can take into account that preservice teachers may be able to incorporate SRL into their future instruction at a more proficient level than they previously believed.

Studies investigating the ways teacher educators support preservice teachers in teaching SRL to their future students point to specific techniques that can be incorporated into teacher education instruction. In an investigation of a successful attempt to help preservice teachers foster elementary children's use of SRL, Perry et al. (2008) found ten conversational patterns that helped university faculty and mentor teachers foster SRL instruction in preservice teachers:

1. Using explicit language for SRL processes
2. Using examples or making suggestions about enhancing SRL
3. Presenting lesson feedback in terms of SRL
4. Asking questions that encouraged thinking about SRL-related processes or metacognitive thinking
5. Modeling what might be said or done to support SRL
6. Prompting transfer from a specific to a general case of SRL
7. Bringing the mentor teacher into the conversation to reinforce practices that promote SRL
8. Reinforcing an SRL practice when someone else introduced it
9. Highlighting SRL practices in written observations
10. Using a discussion protocol that emphasizes SRL

In their interactions during teacher education instruction and in field experiences, the university faculty and mentor teachers found opportunities to embed SRL processes into discussions with the preservice teachers. The topic of SRL was pervasive in the learning experience and was the lens through which learning was analyzed in their coursework. Language for SRL processes was used in initial discussions of methods of teaching, lesson feedback, discussions of field experience implementation, and in reflection experiences. These intentional experiences were built into university instruction and into field experiences similar to the student SRL instruction suggested in the various science and engineering practice instructions in previous chapters.

Preservice teachers who have more developed metacognitive ability themselves may be more open to explicit and scaffolded SRL instruction in their future classrooms (Kramarski & Michalsky, 2010). Therefore, an important antecedent to preservice teacher SRL instruction is to support individual preservice teachers' personal SRL processes. Fortunately, it has been shown that metacognitive scaffolding can support preservice teacher regulation of their own learning (Dembo, 2001; Gordon et al., 2007; Kramarski, 2008; Kramarski & Revach, 2009). The same features of supporting student SRL processes illustrated in earlier chapters can be useful for supporting preservice teacher SRL so that they are open to teaching their own students SRL in the future. Not only can support for preservice teachers in this way foster student SRL instruction, but preservice teacher SRL instruction has also demonstrated a positive impact on preservice teachers' design of lesson plans, classroom performance, creative problem solving, and design of a student-centered classroom (Kramarski & Michalsky, 2010; Kramarski & Revach, 2009).

12.3 Recommendations for Teacher Education Instructional Design

Supporting preservice teacher instruction of science and engineering practices and SRL can go hand in hand. The research of both areas suggests that building strong preservice teacher beliefs for success in both areas can set the stage for fruitful planning and implementation. Preservice teacher research-based instruction on science and engineering practices and SRL can be summarized into three key ideas: beliefs, modeling, and feedback loops.

Before preservice teachers can be proficient in designing and teaching lessons on science and engineering practices and SRL, they should have positive beliefs in their ability to do so and value teaching students these topics. Teacher educators can assess preservice teacher self-efficacy and task value of teaching science and engineering practices and SRL. Teacher educators can then use the preservice teacher responses to have explicit discussions about task value and self-efficacy and the importance thereof. Incorporation of SRL processes in discussions is important to aiding preservice teachers to be comfortable with using SRL supports in their own classrooms in the future.

Research also shows that preservice teachers may not have had first-hand experience of science and engineering practices or SRL as young learners, so it is recommended that teacher educators develop methods in their classes so as to provide preservice teachers opportunities to participate in science and engineering practices and to become aware of their own SRL. Teacher educators should also take the opportunity to incorporate the language of SRL into the experiences so that these underlying processes are revealed to preservice teachers. Doing so may help preservice teachers to integrate the same explicit instruction they received into their designed lessons for their own classrooms. Not only will this reinforce a preservice teacher's own SRL but will open opportunities for preservice teachers to see how to effectively teach science and engineering practices and to support student SRL simultaneously. Preservice teachers can use teacher educator modeling of these skills and translate them for their own future classrooms.

Although explicit instruction is helpful for preservice teachers to incorporate science and engineering practices and SRL supports into their lessons, they need to receive multiple instances of feedback on their own performance in order to be comfortable with and proficient in teaching these skills. Teacher educators can foster an environment that embraces feedback and communication, not necessarily constant evaluation. Feedback seems to be most helpful when it is non-threatening and is used for two-way communication leading to improvement. Setting up preservice teacher goal orientation to lean toward mastery in the course will help them to be more comfortable with facing inevitable failures and receiving constructive criticism. As always, the use of the language of SRL is helpful in revealing underlying skills and processes so that all can learn from the exposure.

12.4 Inservice Teachers Professional Development in Science and Engineering Practices

Research on inservice teacher beliefs in science and engineering practices can be a foundation for professional development in this area. Hang and

Srisawasdi (2021) explored inservice teacher beliefs ($n = 100$) about the importance of the eight science and engineering practices in a survey. They found that inservice teachers rated the practices of explaining and argumentation highest in importance and indicated that they taught this practice once or twice a week. They rated asking questions, defining problems, data collection, and data analysis next in importance and indicated that they taught these practices once or twice a month. Like preservice teachers, the inservice teachers rated modeling as lowest in importance among the practices and indicated that they rarely taught modeling.

There are several approaches to designing professional development, given the knowledge of teacher beliefs in the value of the science and engineering practices. Professional development program design can focus on what is easier for teachers to integrate into their instruction first and lead up to more difficult or less valued topics, or professional development program design can focus on more difficult topics first because that is what is most needed at the time. This decision will likely depend on the relationship the instructor has with the teachers and the needs of the teachers. A benefit of the nature of the topics of the science and engineering practice is that the topics are not distinct and by introducing one practice, an instructor will inevitably need to address a suite of practices.

There have been studies that have shown success in designing professional development (PD) that initially addresses what teachers know well and then working to lesser known topics. Duschl and Bybee (2014) recognized the interconnectedness of practices and made recommendations for the first approach, beginning teacher professional development with a more familiar practice, planning and conducting investigations. They suggest using this familiar practice to bridge to less familiar practices in the instruction. They use the 5E curriculum framework (discussed in Chapter 13 of this book) to set the foundation for the instruction as it allows learners several entry points to learning the topic. Learning to plan and implement investigations in this professional development framework offers an authentic and "messy" science experience for inservice teachers. Duschl and Bybee suggested that these authentic experiences can help teachers to be motivated to learn about less familiar practices such as developing and using models. Merritt et al. (2018) reviewed lessons and teacher reflections, examining how kindergarten and first grade teachers incorporated NGSS scientific and engineering practices during inquiry-based instruction. They found that most of the teachers were able to work with their students on asking questions; planning and carrying out investigations; analyzing and interpreting data, using mathematics and computational thinking; and obtaining, evaluating, and communicating information. However, teachers faced challenges in supporting students in developing their own questions that could be investigated and using data collection strategies that aligned with using mathematics and computational thinking. Christian et al. (2021) explored engineering pedagogical self-efficacy and career awareness of secondary STEM teachers who participated in an engineering education workshop. Their findings indicated that participating teachers significantly improved their confidence in engineering pedagogy, as well as their knowledge of engineering careers and pre-college preparation for post-secondary engineering. After participating in the workshop, teachers reported that they saw engineering as a potentially powerful tool in developing students' critical thinking and problem-solving skills, particularly when integrating the practices of science and engineering with the instruction of disciplinary content. Overall, research demonstrates that approaching PD design by connecting to teachers with a familiar science and engineering practice (SEP) and then expanding to less familiar SEPs can be successful and can inform future PD opportunities.

Other studies took a different approach by examining a professional development program that was focused on science and engineering practices that were less familiar to the inservice teachers. Colclasure et al. (2022) studied the outcomes of the Nebraska STEM Education Conference, a PD program for middle school, high school, and first- and second-year post-secondary STEM teachers that focused on developing and using conceptual models and using mathematics and computational thinking. They found from a survey of forty-five of the PD participants that they had limited prior use of science and engineering practices in their teaching and student interest and learning outcomes were the factors found to be most influential to teachers' use of science and engineering practices. The teachers also reported that their most common barriers to teaching science and engineering practices were limited knowledge, confidence, and resources. Although teachers faced barriers, the Nebraska STEM Education Conference was successful in significantly improving the teachers' confidence and interest to incorporate science and engineering practices. Guzey et al. (2014) studied the effects of a PD for 198 elementary teachers from 43 schools in 17 districts who participated in a yearlong professional development program designed to help integrate the science standards, with a focus on engineering, into their teaching. They found from analyzing lesson plans and student artifacts that the majority of the teachers who participated in the professional development program were able to effectively implement engineering design lessons in their classrooms. The teachers regarded their participation in the lessons that were modeled by the instructors as the most helpful aspect of the professional development program. Brand (2020) studied an interdisciplinary team, consisting of science education and mechanical engineering faculty and doctoral students from each discipline, and science, mathematics, and career and technical curriculum supervisors, who collaborated with middle school science, mathematics, and career and technical education teachers to develop a framework for integrating engineering practices into their curricula. From this collaboration, the teachers saw value in the science and engineering practices and as a result, teachers were motivated to critique and revise their practices. These studies suggest that teachers' success in implementing science and engineering practices in lessons was closely related to the structure of the professional development program, and that the approach of focusing professional development experiences on lesser known science and engineering practices can also be successful.

12.5 Inservice Teachers Professional Development in SRL

There is evidence that teachers who are effective at self-regulating their own skills are more effective at teaching students SRL skills (Kramarski & Kohen, 2017; Kramarski & Michalsky, 2009). Providing professional development for teachers in explicit SRL is extremely powerful because teachers who are knowledgeable in SRL can pass it on to their own students, amplifying the number of self-directed learners. Knowledge about one's own SRL is also beneficial for teachers because SRL assists teachers in navigating daily classroom challenges (Pazhoman & Sarkhosh, 2019), providing effective instruction (Klusmann et al., 2008; Tricarico & Yendol-Hoppey, 2012), and possessing occupational well-being (Klusmann et al., 2008).

Dignath-van Ewijk and van der Werf (2012) conducted a survey study with forty-seven Dutch teachers who taught seventh and eighth grade students. Teachers in the study reported that it was important to engage in SRL strategy instruction when teaching students to be self-regulated learners. However, when asked to define SRL, teachers mainly focused on a constructivist classroom, with little or no integration of supporting strategies. Similarly, Moos and Ringdal (2012) conducted a literature review of SRL in the classroom and found that although the use of SRL processes by learners may be mediated by personal characteristics, a theme that resonated with successful teaching of SRL was teacher beliefs toward student autonomy and the need for explicit and direct instruction on strategies that support student SRL. They also found that at the elementary, middle, and secondary levels, teachers were only implicitly rather than explicitly reinforcing student SRL behaviors. When teachers did use explicit SRL strategy instruction, as found by Spruce and Bol (2015), teachers most frequently encouraged student SRL during the monitoring phase of learning events in their classrooms. However, in the same study, they demonstrated gaps in knowledge around goal setting for a task and self-reflection after a learning event. Therefore, there is a need to support teachers with explicit, developmental SRL strategies for students so that they can build their SRL skills in a deliberate and systematic way. There is room for professional development experiences to help teachers to focus on all of the processes of SRL in an efficient way. Perhaps in establishing a plan for the deliberate development of SRL skills, teachers can have more confidence in their students for building these skills.

12.6 Recommendations for Professional Development Design

Teaching adults is different in some ways and similar in some ways to teaching children. For example, teaching adults is like teaching children if the adult is new to a topic, so it is best to not overwhelm the learner and begin by teaching something that everyone can succeed in. This sets the stage for higher self-efficacy of the learner for the learning task. However, teaching adults requires an instructor to respect the experiences of the learner, because adults know how they need to apply the new knowledge in their classroom context.

Loucks-Horsley et al. (2003) have compiled some of the ways to approach teacher professional development for mathematics and science teachers that reflect the needs that adults have as learners. They recommend organizing the professional development experience based on a needs analysis of the teachers and allowing for teacher input. Not only should the topics in the professional development experience be of interest to the teachers, but they also must align to the teachers' mandated standards. Teachers are less likely to use the new content knowledge and pedagogical content knowledge if the information taught in the professional development experience is not valued or seen as useful. It is also recommended that the professional development experience be experiential. Teachers can then try the same skills that they will be teaching to their students. Having an understanding of what students need to do to learn the concept or skill or practice is helpful when teachers need to intervene to support students.

Recommendations that have come from research reviewed earlier also show that teachers new to a topic should have the same experiences as a learner before they receive experiences as a teacher of the topic. Teachers should have deep content knowledge before they are expected to translate that knowledge into pedagogical content knowledge for their students. Teachers should also have opportunities to reflect on their learning and planning for students.

These recommendations run parallel to the same recommendations for teaching students science and engineering practices supported by SRL. In terms of forethought processes, teachers may learn more when they value the content of the professional development experience. Teachers may learn more efficiently if they set process goals to get them to the outcome goals of the PD. Teachers who have higher self-efficacy for the learning tasks in the professional development experience will tend to take more risks and get more from the learning. Like students in the performance phase of SRL, teachers will be able to identify the key parts of the professional development if the instructor helps them to focus their attention. Likewise, if teachers set goals for learning in the professional development, they can use these goals to metacognitively monitor their learning. Teachers can then participate in self-reflection on their performance in the professional development. Teachers would have a self-reaction to the feedback they receive, whether it is peer review or student outcomes of the lesson, which could then be evaluated. Teachers can then attribute their successes and failures in teaching the content related to professional development and adapt as necessary.

12.7 Example of an Elementary Teacher Professional Development Course on Using Inquiry to Teach Earth Science

This section will explain the application of research in a professional development setting. The purpose of the sixteen-week professional development course was to give teachers who teach students in grades K–5 experiences in learning earth science content as taught in an inquiry format. Once mastered, the K–5 teachers then designed their own inquiry-based earth science lessons that included science and engineering practices. The aim of the professional development course was to provide K–5 teachers with the knowledge and skills necessary to plan and implement student-centered inquiry-based classes. To that end, the course objectives were as follows:

- Extend and strengthen participants' knowledge of earth science content;
- Enhance participants' ability to design and implement inquiry-based lessons that are aligned with state and school division curriculum documents; and
- Emphasize the nature of science, particularly in terms of helping students to think, act, and communicate.

The professional development schedule and topics are found in Table 12.1. There were planned SRL supports across the professional development course. First, teachers participated in forethought. They learned about the goals of the course and set their own goals about learning at the beginning of the course. Key aspects of inquiry were identified, first using teacher experience, filling in any gaps in the information with evidence from videos of expert inquiry teaching. Then teachers participated in the performance phase by accomplishing processes of attention focusing and metacognitive monitoring. Teachers in this phase worked on recognizing what students experience in inquiry and noticing how the lesson plans deliberately set up environments for inquiry learning. After teachers evaluated other lesson plans, they were asked to author their own for their classes. There were multiple attempts at providing feedback on the lesson plan from the instructor and from the peers to hone the teachers' ability to focus their attention on key features. Next, teachers learned more content knowledge and pedagogical content knowledge about other earth science topics by evaluating lesson plans in the last part of the course. Finally, teachers taught two inquiry lesson plans that they designed and reported on the outcome. The class then peer-reviewed the lessons and provided feedback the next time the teacher taught the lessons. Teachers participated in self-reflection on this feedback at the end of the course and provided information to the instructor about how they learned both content knowledge and pedagogical content knowledge during the professional development course.

A study was conducted of elementary teachers' SRL processes while being engaged in the professional development course (Peters-Burton & Botov, 2017). Understanding the teacher SRL processes gave insight into how they learned. It was found that teachers' collective self-efficacy was low for learning how to teach earth science content using inquiry. Although teachers' self-efficacy for teaching earth science using inquiry was low, their task value was high throughout the course. Because they saw value in learning how to teach using inquiry, the teachers persisted in the

Table 12.1 *Topics, goals, and format for professional development on learning how to teach earth science through inquiry for grades K–5 by week*

Weekly topic/learning experiences	Goals	Format for learning experience
Inquiry Nature of science Content map	To identify key aspects of inquiry To identify connections between aspects of inquiry and the nature of science Create a map or webbing of content ideas in earth science for the grade level	Provide anecdotes about ways inquiry has been used in their classrooms Provide anecdotes about ways the nature of science has been used in their classroom Watch videos of inquiry and nature of science being taught at an expert level (outcome goal) Use evidence in video to find areas for improvement and set goals Use learning standards to map out content and skill learning for students over the school year Answer questions about forethought processes
K–1 inquiry lesson Earth science strand	To engage in a K–1 inquiry lesson on earth science as a student To reflect on learning through inquiry To create or adapt a K–1 level lesson on earth science to use inquiry methods	Read lesson plans for K–1 inquiry lesson on earth science topic Participate in the lesson in the role of a student Collaboratively create or adapt a K–1 level lesson on earth science to use inquiry methods and peer review
Grade 2–3 inquiry lesson Earth science strand	To engage in a grade 2–3 inquiry lesson on earth science as a student To reflect on learning through inquiry To create or adapt a 2–3 grade level lesson on earth science to use inquiry methods	Read lesson plans for grade 2–3 inquiry lesson on earth science topic Participate in the lesson in the role of a student Collaboratively create or adapt a grade 2–3 level lesson on earth science to use inquiry methods and peer review
Grade 4–5 inquiry lesson Earth science strand	To engage in a grade 4–5 inquiry lesson on earth science as a student	Read lesson plans for grade 4–5 inquiry lesson on earth science topic

Topic	Objective	Activity
	To reflect on learning through inquiry	Participate in the lesson in the role of a student
	To create or adapt a grade 4–5 level lesson on earth science to use inquiry methods	Collaboratively create or adapt a grade 4–5 level lesson on earth science to use inquiry methods and peer review
		Teacher-authored Lesson Plan #1 due for instructor feedback based on content map
Reporting out of inquiry Barriers and benefits	To compose a list of barriers and benefits of inquiry in elementary school science	From notes taken during professional development, collaboratively develop a list of benefits and barriers to teaching inquiry in elementary school science
		Report out benefits and barriers to make a class list and discuss until there is consensus
Lesson plan evaluation – Interrelationships of Earth/Space Systems K–5	To evaluate an inquiry lesson plan on a topic (Interrelationships of Earth/Space Systems) for content accuracy and pedagogical content knowledge for potential effectiveness	Given a lesson on interrelationships of Earth/Space systems, evaluate the lesson for content accuracy and pedagogical content knowledge for potential effectiveness and discuss as a group
Lesson plan evaluation – Earth's patterns, cycles and change	To evaluate an inquiry lesson plan on a topic (Earth's patterns, cycles, and change) for content accuracy and pedagogical content knowledge for potential effectiveness	Given a lesson on Earth's patterns, cycles, and change, evaluate the lesson for content accuracy and pedagogical content knowledge for potential effectiveness and discuss as a group
Lesson plan evaluation – Earth resources	To evaluate an inquiry lesson plan on a topic (Earth resources) for content accuracy and pedagogical content knowledge for potential effectiveness	Teacher-authored Lesson Plan #2 due for instructor feedback based on content map
		Given a lesson on Earth resources, evaluate the lesson for content accuracy and pedagogical content knowledge for potential effectiveness and discuss as a group
Peer review of Lesson Plan #1	To offer constructive criticism on lesson plan design for elements of inquiry, the nature of science, and content accuracy	Answer questions about Performance Phase of SRL

Table 12.1 (*cont.*)

Weekly topic/learning experiences	Goals	Format for learning experience
	To answer questions about one's own performance processes when learning to teach earth science using inquiry	Lesson Plan #1 revision due Peer review of Lesson Plan #1
Peer review of Lesson Plan #2	To offer constructive criticism on lesson plan design for elements of inquiry, the nature of science, and content accuracy	Lesson Plan #2 revision due Peer review of Lesson Plan #2
Present lesson plans to group (takes several weeks)	To offer constructive criticism on lesson plan design for elements of inquiry, the nature of science, and content accuracy	Lesson plan presentations of lessons #1 and #2
Celebration Next steps Evaluation of the course	To reflect on learning about earth science topics To reflect on learning about teaching earth science using inquiry To offer constructive criticism on teacher professional development	Reflect and discuss personal learning about earth science topics Reflect and discuss learning about pedagogical content knowledge for earth science inquiry Answer questions about self-reflection phase of SRL

professional development course, even when they were overwhelmed or had failures when they taught the content. Teachers reported that they were knowledgeable about how they focused their attention, so they were able to metacognitively monitor their learning. By the end of the 16-week professional development course, all of the teachers reported that they were satisfied with their performance.

12.8 Example of Secondary Teacher Professional Development Course in Scientific Argumentation

This section describes the application of the research findings about in-service teachers learning about science and engineering practices and SRL. The topic of the professional development course was scientific argumentation and the goals were twofold: (a) to have teachers master the Claims-Evidence-Reasoning format and use of argumentation in science and (b) to have teachers develop their own lessons that feature the use of arguments with evidence in science. The first part of the professional development course, which totaled twenty-four hours of instruction, focused on teachers' own learning about argumentation and discourse in science. The second part of the professional development course, which totaled another twenty-four hours of instruction, focused on how teachers can support students to learn how to evaluate scientific arguments and develop arguments from evidence. The professional development schedule and topics are found in Table 12.2. By using SRL as a framework, details about teacher learning processes for each of the three phases of two learning cycles of teachers (learning and pedagogical content knowledge) were assessed (Figure 12.1).

A study was done on this professional development program and participating teachers responded to Forethought, Performance, and Self-Reflection forms during two different learning cycles: the cycle of learning about argumentation and the cycle about teaching argumentation in science (Peters-Burton, Goffena, & Stehle, 2022). The results of the study reinforced the idea that teachers have deeper knowledge when they first learn content knowledge and then learn about pedagogical content knowledge. The teachers reported that one of the most helpful activities in the professional development course was the collaborative construction of a rubric that helped them understand their proficiency in understanding the practice of argumentation. The rubric was useful because it articulated the smaller process goals that accumulate to the outcome goal. Once teachers mastered process goals with the aid of the rubric, then they were able to be efficient learners and only needed the outcome goal to guide their learning.

Table 12.2 *List of objectives and activities from a thirty-two-hour professional development program focused on learning argumentation and teaching argumentation in science*

Objectives	Activities
Before the PD program began, teachers were given a research question and data with an explanation of how the data were collected. Teachers were instructed to write an argument to answer the research question. At strategic times in the PD, teachers revised their argument after they learned about that concept.	
<ul><li>Articulate the goals of the PD</li><li>Define argumentation</li><li>Evaluate your argument</li><li>Develop evidence and reasoning for an argument</li></ul>	Explain PD expectations SRL forethought questions What is argumentation in science?<ul><li>Definitions from research</li><li>NGSS – vertical articulation</li></ul>Why argumentation?<ul><li>Think/pair/share</li><li>Overall discussion</li></ul>First reaction to teacher-authored argument<ul><li>How does it fit in scientific practices?</li><li>Is your argument convincing?</li></ul>Basic argumentation<ul><li>Elements of an argument</li></ul>Did Chester or Piobar eat the butter? activity<ul><li>Evidence and warrant</li></ul>Second reaction to your constructed argument<ul><li>Evidence and rule discussion</li><li>Prepare for tomorrow</li></ul>
<ul><li>Evaluate a constructed argument</li><li>Determine factors in inferential distance</li><li>Determine factors in developing warrants</li></ul>	Discussion of rationale for teaching argumentation and parts of an argument (analyze a video) Analyze a written argument<ul><li>Underline and highlight</li><li>Discussion of parts of argument</li><li>Is the argument convincing?</li></ul>Inferential distance<ul><li>Instruction</li><li>Evaluate your argument</li><li>Evaluate other teachers' arguments</li></ul>Developing warrants<ul><li>With mascots</li><li>With scientific arguments</li><li>With your arguments</li></ul>SRL performance assessment
<ul><li>Determine reliable resources</li><li>Conduct peer review of arguments</li><li>Construct rebuttals of arguments</li></ul>	Discussion of rationale for teaching argumentation and parts of an argument (analyze a video) Write your argument for these questions. Cite resources when possible

Table 12.2 (cont.)

Objectives	Activities
	- What qualifies as science? At what point is an argument scientific? What criteria are upheld that make a claim scientific? - Discuss arguments in small groups - Report out for big group - What did you learn about arguments? Rebuttals - Why the rebuttal – opposing arguments make your stronger - Sample argument essay outline Rebuttals continued - Highlighting exercise - Argument essay outline for other scientist argument - Argument essay outline for your argument Reviewing key information - Parts of an argument - Inferential distance - Warrants - Rebuttal

Review argumentation by developing a rubric for evaluating written arguments
- Revisit developing an argument from data
- Develop general rubric for developing arguments from data
- Clarify your goal for altering your lesson plans
	What is argumentation in science?
	- Definitions from research
	- Evaluating the argument from data – developing content knowledge article
	Considering collaborative groups
	- Review your three lesson plans and be ready to talk about the following five things for each lesson plan
	  o Grade level
	  o Topic
	  o Focusing on verbal or written or both
	  o Focusing on evaluating an argument or developing an argument
	  o What is your goal for this lesson plan by the end of the PD?
	Creating a rubric for evaluating written arguments
	- Overview of rubrics handout
	- Share Sampson rubric
	- Create a general rubric as a group
	- Overview of constructed response items (essays)
	Get into groups where at least one member has a lesson plan that focuses on evaluating arguments
	- Look over your lesson plans that focus on evaluating arguments

Table 12.2 (cont.)

Objectives	Activities
	• Revise question and rubrics as a team Creating arguments from data • Ice cube data set • What arguments can be made? Are they good? • Create a rubric of our own Get into groups (or create new ones) to review lessons creating arguments from data • Work in teams to revise

Explain model-evidence-link (MEL) diagrams
• Apply rubric to MEL diagrams
• Explain what we know about how students learn argumentation
• Develop scaffolds to build student collaborations
• Evaluating verbal arguments
• Apply findings from educational research to your lesson plans
	Sample activities on written argumentation
	• Model-evidence link diagrams
	  o Overview
	  o Pre-task ranking (together)
	  o Creating MEL diagrams (small groups)
	  o Peer review (across groups)
	  o Reviewing rubric with arguments from MEL diagrams
	What we know about teaching argumentation in the science classroom from research
	• Learning progressions
	  • https://www.youtube.com/watch?v=9tM-G_dZnvs
	• Note on lesson plans areas to work on based on educational research findings
	Building scaffolds between practice and research
	• In groups, review your lesson plans for appropriate scaffolds from the research literature
	• If you don't know what to do yet, just note where you feel there is a gap
	• Do the performance SRL microanalysis questions
	Evaluating verbal arguments
	• Watch video of lab group argumentation
	• Create a general verbal argument rubric (sample)
	Work on Lesson plans
	• Verbal arguments – put them in to support written arguments

Describe SRL theory and ways to assess student SRL orientation • Describe the coaching strategy of Metacognitive Prompts • Apply findings from educational research to your lesson plans	Metacognitive prompts coaching strategy Work on argumentation prompts Work on lessons

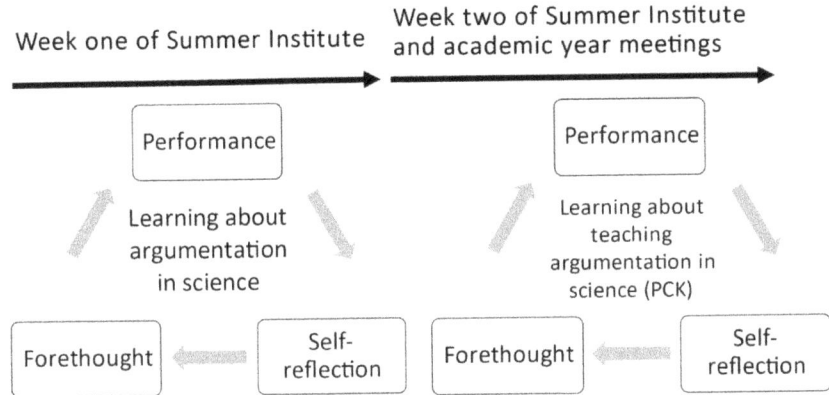

Figure 12.1 Progression of instruction for scientific argumentation professional development

Not only did the teachers learn about the science and engineering practice efficiently, they were able to articulate their SRL process on the forms. The SRL forethought, performance, and self-evaluation questions helped the professional development instructor to change course based on the teachers' needs. For example, at the beginning of the professional development experience when teachers were supposed to focus on their own learning, they reported goals that were focused on their students' learning, rather than their own learning. Because the goals were incongruent with the professional development experience, the instructor was able to have a timely discussion so that teachers' goals could be realigned. Although the SRL processes forms took some time in the professional development experience to fill out, ultimately the time spent was efficient for learning about argumentation and learning how to teach argumentation.

CHAPTER 13

Planning Lessons with Embedded Self-Regulated Learning Using the 5E Format

The study of teacher lesson planning is a thriving area of research in education, and understanding the process and products of lesson planning is needed to plan an effective teacher education. This chapter will focus on how self-regulated learning strategies can be embedded through the use of a familiar tool, the 5E lesson format. An example of a physics lesson focused on investigating the movement of a pendulum with embedded science and engineering practices and SRL processes is explained at the end of the chapter.

The 5E model of curriculum design has been widely used by elementary and secondary science teachers since the last twenty years (Bybee et al., 2006). The 5E model uses the components of Engagement, Exploration, Explanation, Elaboration, and Evaluation as the steps of inquiry for students within a lesson plan that often features investigations. The 5E model is useful for framing a lesson as a student-centered inquiry approach, and it is useful for embedding science and engineering practices in lesson plans. The use of the 5E lesson planning model leads to the explicit design of science and engineering practices. Additionally, because the 5E model is a format for science and engineering problem solving, the model shares a symbiotic relationship with SRL.

13.1 Elements of the 5E Model of Curriculum Design

There are five elements in the 5E model of curriculum design: Engagement, Exploration, Explanation, Elaboration, and Evaluation. These elements are intentionally designed to introduce students to new concepts through the use of connections to prior knowledge and the intentional building of knowledge and skills. The elements of the 5E model are meant to be iterative. For example, a teacher may begin with engagement, move to exploration, and then explanation. At the end of explanation, the teacher may realize the students need more experience in exploration and provide more instruction in the exploration element. Similarly, if students are involved in the explanation phase of the curriculum and a new concept or skill emerges, then the teacher will likely go back to engagement or exploration to give the students a chance to interact more fully with the new concept or skill. Although a summative evaluation is intended to be administered at the end of the 5E cycle, formative evaluations are also needed at the end of each element in the cycle so that a teacher is informed about student mastery before they move on to the next element of the 5E design. In the next section each element of the 5E model will be explained in terms of what the teacher does during this element and what the student does during this element. It will also include science and engineering practices and SRL processes that are compatible with the element.

13.1.1 Engagement

The purpose of engagement in the 5E model is to frame the overarching ideas, generate motivation, and connect prior knowledge to the topic under study. During engagement, the teacher sets up a learning environment that gives students clues to the types of phenomena they will investigate along with possible connections to prior knowledge and experience. The purpose of an engagement environment is to motivate, create interest, and foster task value in the students. The teacher helps students to ask questions about the phenomena and record what students ask about the phenomena, but should not yet answer student questions. During engagement, students should try to make sense of what the teacher is introducing and ask questions to clarify ideas. Students should try to determine what they currently know about the phenomena and record what they may need to know in the future.

The science and engineering practices that are most compatible with 5E engagement are asking questions and defining problems, and developing and using models. As students are initially engaging with the phenomena, they can simultaneously develop a model of what they currently know, leaving space for what they may be learning in the future. Asking questions and defining problems are also well-suited to be carried out in the engagement element because students can begin forming their questions as they are introduced to the phenomena. Asking questions and defining problems may be refined in the next phase, exploration as well.

The SRL processes in the forethought phase are well-matched to activities during the engagement element. Students can begin to set goals and strategically plan how to reach those goals when they are introduced to a topic, and then can refine goals and strategic plans as they move on to exploration. Students will likely feel a particular self-efficacy and have a goal orientation for learning a topic when they are introduced during the engagement phase. It is here where teachers are also building task value for learning the topic, and in conjunction, teachers can make opportunities to build self-efficacy for the topic as well.

13.1.2 Exploration

The purpose of the exploration element in the 5E model is to introduce students to new concepts and skills, refine ideas from engagement, and reflect on relationships and concepts in the phenomena. Teachers during the exploration phase act as facilitators for student questioning and refinement of ideas. They listen to student discussions and encourage reflection on key concepts and skills. In a sense, teachers are preparing students to embark on the deeper learning for the next element, explanation. Students should conduct activities regarding the phenomena, make predictions about variables and relationships, share refined ideas, record patterns they observe, and make conjectures about alternative ideas that could be possible during exploration.

The science and engineering practices that are compatible with exploration are developing and using conceptual models, planning and carrying out investigations (on a small scale), using mathematical and computational thinking, and evaluating and communicating information. In the exploration phase, students can refine the conceptual models they drafted during engagement. Because they are performing small tests with the phenomena to find general patterns, they are planning and carrying out small investigations as they explore. During exploration, students are using computational thinking such as identifying patterns and decomposition as well as being engaged with scientific information.

The SRL processes that match well with the 5E exploration element are forethought and performance. Students may be gaining self-efficacy and task value for the topic as they explore deeper into the ideas as they engage in forethought. Exploration brings out performance processes of attention focusing and metacognitive monitoring because students are beginning to refine their ideas with a realization of material that is extraneous information.

13.1.3 Explanation

The purpose of the explanation element of the 5E model is for students to begin to use formal language to make connections between prior knowledge and the phenomena, and to communicate new ideas about the phenomena. This is where deeper learning of the key concepts can happen. The role of the teacher in explanation is to encourage students to explain their newfound understandings of the phenomena and to ask clarifying questions and justifications for students to dig deeper into understanding the phenomena. There is often a misconception that this phase is where teachers should be explaining the phenomena, but it is where teachers should allow students to explain their understanding of the phenomena. Students, during explanation, should use their previous understandings and findings to communicate their view of the phenomena variables, relationships, and mechanisms, if possible. Students should listen to each other's explanation and attempt to refine their understanding based on peer explanation. Students should also provide answers to the teacher's clarification and justification questions to the best of their ability.

In terms of relevant science and engineering practices, using conceptual models comes into play during explanation. It is during this time that students build formal detail into their conceptual models. Students may also develop and carry out investigations on big ideas related to the phenomena in order to come up with those deeper ideas to add detail to their conceptual model. It is also likely that students will be analyzing and interpreting data, constructing explanations or engaging in argumentation, using mathematical and computational thinking, and evaluating and communicating information during explanation. In order to make sense of their investigations to form them into deeper ideas, students will need to use mathematical and computational thinking to analyze and interpret the new ideas they have developed. In the same way, students will need to construct explanations, or if they are trying to convince others of their unique idea, will have to engage in argumentation. During this whole process, students will be consuming information and will need to make decisions on what is valid and able to support or refute their thinking.

The SRL processes that match well with explanation are performance and self-reflection. During explanation students are forming formal ideas and are engaging with attention focusing on the key information. As students explain their ideas, they are receiving feedback from the teacher and peers (or an external audience) and are self-reacting and self-evaluating the feedback. They likely are attributing their successes and failures in communicating the information as well as adapting their thinking to refine their model of the phenomena.

13.1.4 Elaboration

The purpose of the elaboration element in the 5E model is to extend the learning from explanation into a new situation and to continue to use formal language and ideas about the phenomena in a new context. During elaboration, teachers set up a learning environment that gives students a new context in which to apply their learning. In doing so, teachers help students make connections between their learning and the new context as well as encourage the use of formal language in student explanations or arguments. During elaboration, students make connections between the learning they did during explanation and the new context while making attempts to use the formal language. They should probe their new knowledge to make sense in this new context and refine their thinking if there is a mismatch.

The science and engineering practices that are compatible with elaboration are using conceptual models, constructing explanations, using arguments with evidence, and evaluating and communicating information. In certain circumstances, students may be conducting additional investigations, so then they would also employ asking questions and designing solutions, developing and carrying out investigations, and analyzing and interpreting data. The elaboration element of the lesson gives students an excellent opportunity to further refine their conceptual model by placing the same phenomena in a different context to see if their model is feasible in this context. Students will also be constructing explanations of the phenomena in the new context as well as using argumentation to explain the variables, relationships, and possibly mechanisms in this new context.

The SRL processes of performance and self-reflection align well with elaboration because students are focusing their attention and metacognitively monitoring when they are using their learning during explanation as a template for the new context in elaboration. Students are also receiving feedback about the utility of their conceptual model for explaining the phenomena in this new context, so they are employing the processes of self-reaction, self-evaluation, attribution, and adaptivity during elaboration.

13.1.5 Evaluation

The purpose of the evaluation element in the 5E model is for students to demonstrate understanding of the goals of the lesson and assess the quality of learning during a particular time frame. The role of the teacher is to develop assessments that will allow students to show that they met the learning goals of the lesson and to inform students of the quality of their learning. The role of the student in the evaluation element of the 5E model is to demonstrate an understanding of the phenomena to a level of expectation established by the teacher.

Depending on the timing and purpose of the evaluation, students may be exhibiting any of the eight science and engineering practices, since the purpose of the evaluation is to assess each of the practices.

The SRL process that is most aligned with evaluation is self-reflection. Students receive feedback on their performance during the evaluation, from which they will self-react, self-evaluate, make attributions, and adapt. Teachers also receive feedback on how well students met the learning goals given the learning environment, from which teachers may iteratively return to a different element of the 5E model to reteach ideas.

13.2 Connections Between the 5E Model of Curriculum Design and SRL

There is a strong link in the research that connects the concepts of problem solving and SRL. When learners begin to sort out what they understand about their own learning through observing and reflecting on their SRL processes, they begin to solve problems with their learning processes by examining processes that work and do not work. Learners who strategically apply cognitive, metacognitive, and behavioral strategies tended to perform better than their peers who were less strategic (Efklides, 2011; Lucangeli & Cabrele, 2006). Due to this overlap, developing lessons using the 5E model prepares a foundation for embedding SRL into a lesson. Table 13.1 displays the ways that the 5E model of curriculum design encourages student problem solving and how the elements of the 5E model overlap with the SRL phases of forethought, performance, and self-reflection.

13.3 Example of 5E Lesson Plan with Task Analysis for Science and Engineering Practices and SRL Processes

In this section, a 5E lesson plan on pendulums for secondary students is presented. After the lesson is explained, a task analysis tool is used to demonstrate where science and engineering practices and SRL processes can be integrated into the lesson. The purpose of using the task analysis tool is to decompose the lesson into learning tasks so that advantageous

Table 13.1 Connections between the 5E model of curriculum design and SRL phases

	Forethought	Performance	Self-Reflection
Engagement	Promotes student interest Make connections between past and present learning experiences Organize student thinking toward the learning outcomes		
Exploration	Provide students with common base of experiences Elicit relevant content knowledge and process skills	Generate new ideas related to content Explore possibilities Design and conduct a preliminary investigation	
Explanation		Explain understanding of the concept or skill	
Elaboration		Apply understanding to new contexts or more complex situations	
Evaluation			Feedback from teacher or curriculum guides student toward deeper understanding Evaluate the utility of currently understood concepts and skills in new or more complex situations Self-assess the links between prior and new knowledge and skills Consider feedback from teacher or curriculum in their ability to acquire new knowledge and skills Catalog what worked and what did not work in attempts to generate new knowledge and skills from future investigations

points are located for the integration of science and engineering practices and SRL processes.

13.4 Pendulums Lesson

13.4.1 Objectives

At the conclusion of the lesson, students will be able to

1. Describe how and why to change one variable at a time.
2. Relate concepts of motion, direction, and speed by explaining the action of pendulums.

13.4.2 Materials

String
Scissors
Tape
Meter stick
Protractor
Various weights to use for pendulum bobs such as washers

13.4.3 Engagement

1. Play at least four video clips of action movies where the actor swings into a car or object on the move, or other actions in movies where a pendulum is involved.
2. Ask the students what they notice about the video, guiding them to observe a swinging person or object at the end of a rope or pole. Replay the videos if necessary.
3. Ask the students, "How do you think they get the timing just right in those videos?" and allow students to form small groups to jot down ideas.
4. Ask students using their ideas from the video to draw a conceptual model of an object on the end of the rope that has a fixed point at the top. Ask students to do their best to label any variables on this model. Check student conceptual models as an evaluation before moving to Exploration.

13.4.4 Exploration

1. Put students in groups of four. Each student has an assigned role. The principal investigator is the person responsible for ensuring that everyone else in the group knows the goals and process of the investigation. Materials manager is responsible for ensuring that everyone in the group has a chance to manipulate the materials in the investigation. The recorder is responsible for everyone in the group having the correct information and data from the investigation. The reporter is responsible for asking questions and communicating information on the group's behalf. The purpose of the roles is to create interdependence among the group members for the investigation.
2. One student, the materials manager, goes to the supply table to pick up a length of string, scissors, tape, and washers of various sizes and weights.
3. Each group is directed to use these materials to:
   - Construct a pendulum
   - Hang the pendulum so that it swings freely from a pencil taped to the surface of the desk
   - Count the number of swings of the pendulum in fifteen seconds.
4. The recorder in each group records the result in a class chart. Students are asked to examine the class data. Each group's numbers will be different and there should be some discussion by students as to why this is so.
5. Groups may be encouraged to repeat their experiment to determine if there is an error in the data. Again, a discussion should ensue to suggest why different pendulums will swing differently. Students should be modifying their conceptual model when new and different information emerges from the discussion. Check their conceptual model as an evaluation before moving to Explanation.
6. Write suggestions on the board as students offer them, such as the weight of the washer, the length of the string, the diameter of the washer, and how high the student starting the pendulum held the washer.

13.4.5 Explanation

1. Each group is asked to design an investigation to test one of the suggestions. The original pendulums are collected and will be used

Planning Lessons with Embedded Self-Regulated Learning   237

later. Newly constructed pendulums are to be used in their new investigations. Before groups begin, check their investigation design, guiding them to change only the one variable that pertains to the suggestion they are testing. For example, one group keeps the string at the same length but attaches washers of different diameters and starts the swing at exactly the same place. Another group uses one piece of string and one washer, but starts the swing at higher and higher places on an arc. A third group cuts pieces of string of different lengths, but uses one washer and starts the swing at the same place each time. Each group is asked to make predictions about the outcome of their specific investigation before they begin to record data.

2. Each group shares their procedure and the data they collected with the rest of the class. Collectively analyze and interpret the data. The class should conclude that the difference in the number of swings that a pendulum makes is due to the different lengths of the string.

3. Students should modify their conceptual model if needed. Note: It is usually surprising to students that weight and size of washers are not factors in pendulum speed. Use their questions during this discussion to segue into additional concepts at the end of this lesson.

4. Write on the board a series of numbers across the top ranging from 7 to 17. The recorder from each group is instructed to tape the group's original (first) pendulum to the board under the number representing the number of swings the pendulum made in 15 seconds. When all pendulums are hung, the students are asked to interpret the results. After discussion, students should conclude that the number of swings in a fixed time increases in a regular manner as the length of the string gets shorter.

5. If the original pendulums hanging on the board were constructed with swings between 9 and 15, for example, but there were no pendulums with swings of 12, then have student groups construct a pendulum with 12 swings per fifteen seconds. Students will measure and cut and measure and test their pendulums to achieve success at this task. Now ask each student to make a drawing in their journals to plot the data on the relationship between the length of the string and the number of swings. Most students will draw the board with pendulums of different lengths, but some will draw charts or graphs of these data.

6. As an evaluation, ask students to communicate how they now think people in the movies were able to swing with such accuracy. Explain that they must construct an argument using the evidence from their investigation.

13.4.6 Elaboration

1. Challenge students to find examples of pendulums at home and in their neighborhoods over the next few days.
2. Discuss graphing to have all students move from their journal drawings to lines, then to points on a graph, and finally to a complete graph. Finally, ask each student to use their graph to make a pendulum that will swing a number of times (specified by you).
3. Ask students to diagram the pendulum and its motion and to indicate on the diagram where the pendulum is in motion and where kinetic energy is reducing and potential energy increasing and vice versa. Ask students to consider what forces are acting up on the string. Through questioning, you will likely get them to identify gravity, air resistance, and tension of the string. Guide students to determine that air resistance is negligible and that the tension of the string is largely constant so therefore, would be unlikely to play a role in determining the number of swings. That leaves only two variables that seem to be impacting the pendulum.
4. Introduce the equation for period of pendulum. Explain to students that physicists have determined this equation for explaining the periodicity of a pendulum. Ask them if their data supports this equation and that the only variable involved in the formula is the length of the string. Ask them if it also matches with the number of swings on their graphs.
5. Ask students to relate this result to previous classes where it was demonstrated that the two objects, regardless of mass will fall at the same rate. Students should be able to explain that the length of the string influences the number of swings because it restrains how long the pendulum can stay in motion on one side of the arc.

13.5 Task Analysis Tool

The purpose of a task analysis tool is to demonstrate a short-hand version of the lesson so that ideas can be integrated into a lesson. For example, the task analysis tool seen in Table 13.2 integrates science and engineering practices and SRL processes into the lesson. The task analysis tool can then be used as a guide to return to the lesson plan so that teachers can build in explicit language in order to model the science and engineering practices and build student SRL skills in a deliberate way.

Table 13.2 *Task analysis of pendulums investigation*

Teacher does	Student does	Science and engineering practice	Student SRL process
Engagement			
Show videos of movie clips with pendulum motion in them	Notice common characteristics in the motion on the videos	Asking questions	Self-efficacy
	Make conjectures about timing of swinging object in video	Asking questions	Self-efficacy Task value Goal orientation
	Draft conceptual model of swinging object	Developing and using conceptual models	Goal setting Strategic planning
Exploration			
Put students in groups, assign and explain roles, distribute materials	Construct pendulum and do preliminary tests	Planning and carrying out investigations	Self-efficacy Task value Goal orientation Goal setting Strategic planning
	Report results	Planning and carrying out investigations	Attention focusing Metacognitive monitoring
	Repeat tests if necessary	Planning and carrying out investigations	Attention focusing Metacognitive monitoring
Compile suggestions for explanation	Refine conceptual model	Using conceptual models	Attention focusing Metacognitive monitoring
Explanation			
Coordinate research questions for groups	Plan investigation of period of a pendulum	Planning and carrying out investigations	Attention focusing Metacognitive monitoring
	Conduct investigation	Planning and carrying out investigations	Attention focusing Metacognitive monitoring

Table 13.2 (cont.)

Teacher does	Student does	Science and engineering practice	Student SRL process
Support analysis and interpretation of data	Share procedure and results	Analyzing and interpreting data Constructing explanations	Attention focusing Metacognitive monitoring Attention focusing Metacognitive monitoring
	Analyze and interpret data	Analyzing and interpreting data	Attention focusing Metacognitive monitoring
Facilitate display of original pendulums	Draw visualization to communicate findings	Analyzing and interpreting data	Attention focusing Metacognitive monitoring
	Construct an argument to explain the timing of swinging objects in movie clips	Using arguments with evidence	Attention focusing Metacognitive monitoring
Elaboration			
	Find examples of pendulums in daily life	Using computational thinking	Attention focusing Metacognitive monitoring
Facilitate refinement of graph of length versus period	Add data and refine graph of length versus period of a pendulum	Analyzing and interpreting data	Self-reaction Self-evaluation
	Extend conceptual model to forces and energy on the pendulum	Using conceptual models	Self-reaction Self-evaluation Attributions Adaptation
Introduce mathematical model of pendulum	Interpret fit of data with mathematical model	Analyzing and interpreting data Using conceptual models	Self-reaction Self-evaluation Attributions Adaptation
	Refine conceptual model with variables, relationships, and mechanisms	Using conceptual models Constructing explanations	Self-reaction Self-evaluation Attributions Adaptation

CHAPTER 14

Research Designs for Examining Science and Engineering Practices and SRL

This chapter explains research designs for potential research of student, preservice teacher, and inservice teacher teaching and mastering of self-regulated learning processes while engaging in science and engineering practices. Sample research designs using Maxwell's (2012) research design framework will be presented for studies involving case study design, comparative design, and mixed methods parallel design. This chapter assumes a basic understanding of how to design educational research and is intended to be instructive for putting these types of designs into the context of student self-regulated learning while engaging in science and engineering practices.

The purpose of Maxwell's (2012) research design framework is to show all of the key decisions on one page so that a researcher can balance the interactions of the design choices. The choices in the design framework include the goals of the research, the conceptual framework of the research, the research question(s), the methods of the research, and ways to enhance the validity of the research (see Figure 14.1 for the design diagram). In the goals section, researchers explain why they are attempting the study. The goals can explain how the research will fill a gap or how it may add to the body of literature. The conceptual framework section articulates theories from which the research is grounded or ideas that are connected in a model created by the researcher. The conceptual framework drives the research and helps to situate the findings. The methods section explains how the data will be gathered and lists measures or instruments that may be employed. The validity section explains what the researcher will do to minimize bias and maximize trustworthiness of the research question(s), data collection, and findings.

A researcher can begin considering factors for their research design by filling in the boxes from this diagram in any order, and they do not need to proceed linearly to note ideas for each factor. However, the research question(s) are in the center intentionally because they ground the research and

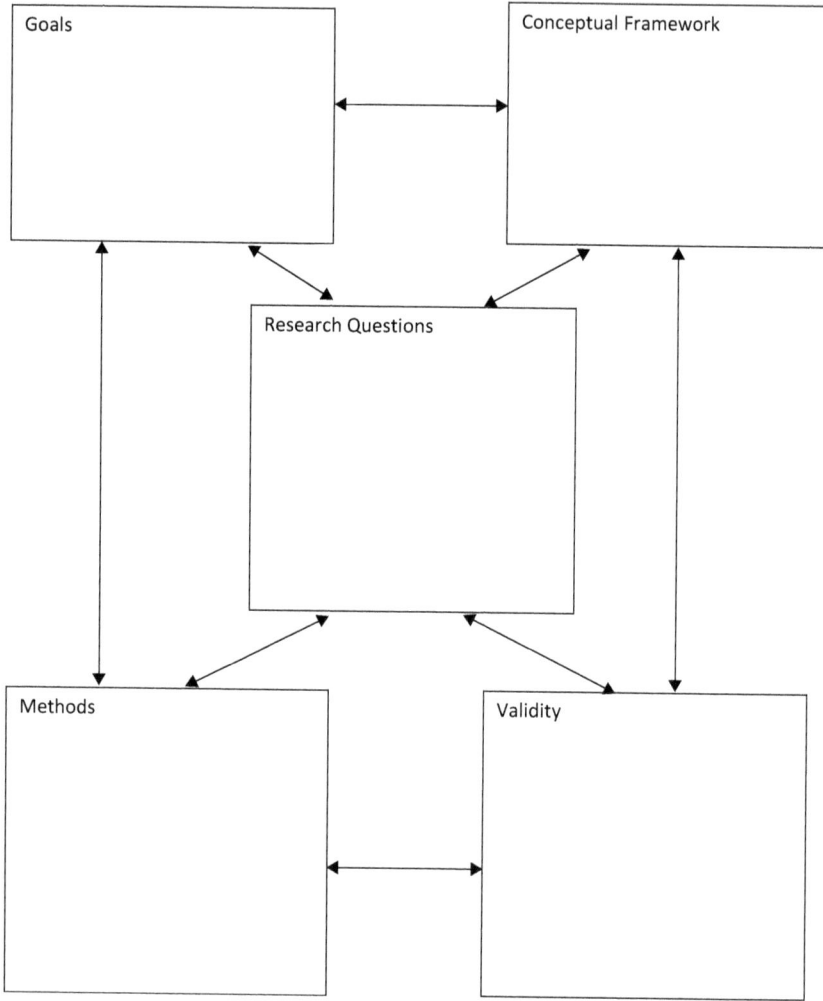

Figure 14.1 Maxwell's research design diagram

may change based on the development of ideas from the other factors. The research question(s) influence all of the other decisions in the design, and depend upon the goals, methods, conceptual framework, and validity checks that are part of the research design. Often a researcher will begin with the goals and conceptual framework in order to find an area of research that is unique and contributory to the field. Then the research question can be formed in service of the area of need. The goals of the research influence the methods and the conceptual framework, and the conceptual framework, in turn, influences the goals and the validity checks. The validity checks influence and are influenced by the conceptual framework and the methods, and the methods influence goals and validity checks.

In the next section, sample research designs focused on exploring how preservice teachers, inservice teachers, and students understand and use SRL to teach science and engineering practices are explained. The case study will be used to plan for research that may explain preservice teachers' understanding of planning lessons with SRL on the topic of science and engineering practice. A quantitative comparative design will be used to plan for research that may decipher any differences or similarities in two groups of students, one group that has explicit SRL support for learning science and engineering practices and another one that has implicit SRL support for learning science and engineering practices. Finally, a research design for a parallel mixed methods study will be explained to plan for investigating inservice teacher learning during a professional development experience involving SRL and science and engineering practices. The research designs presented are broad and overarching. They are intended to serve as ideas from which researchers can fill in details for implementation. The research designs are not intended to be detailed enough to be carried out without further fleshing out.

14.1 Case Study of a Group of Preservice Teachers Learning about SRL for Science and Engineering Practice Instruction

Case studies are useful to explain a particular situation in great detail (Yin, 2003). When focusing on a small group or on individuals, the researcher has the opportunity to gather very detailed data and look very closely at phenomena that are of interest and those that can emerge from the data. In case study design, a researcher must set boundaries for the case, which helps a researcher to decide how much data to collect. Another benefit of case study is that a researcher can explore both processes and outcomes of the participants with qualitative methods. Qualitative methods also have the benefit that the data will be obtained verbatim from the participants rather than asking the participants to interpret an external framework. One of the limitations of case study is that the findings of the study may not transfer to all situations for preservice teachers. However, there are usually parts of the findings that are transferrable to many different situations. As seen in Figure 14.2, the study will be grounded by the following research questions:

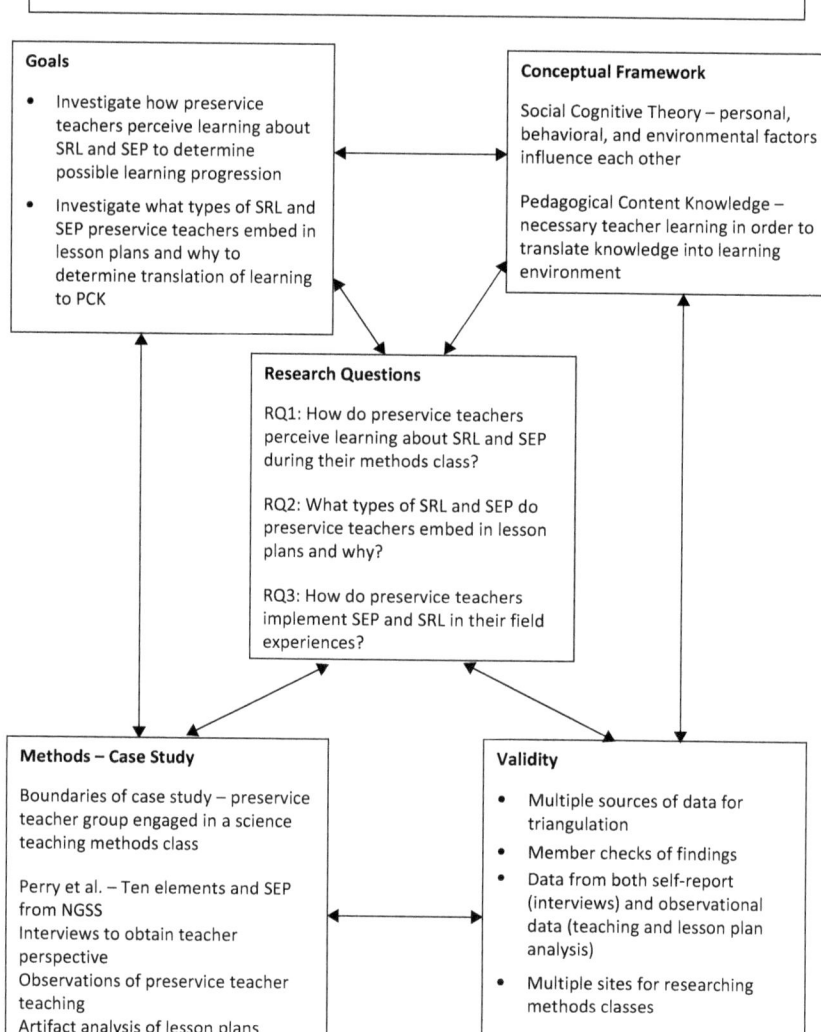

Figure 14.2 Research design diagram for case study

RQ1: How do preservice teachers perceive learning about SRL and science and engineering practices during their methods class?
RQ2: What types of SRL and science and engineering practices do preservice teachers embed in lesson plans and why?

RQ3: How do preservice teachers implement science and engineering practices and SRL in their field experiences?

The goals of the sample case study are to understand how preserve teachers perceive advanced techniques of teaching science and engineering practices and self-regulated learning processes, what methods preservice teachers may feel comfortable adopting, and how preservice teachers plan to teach these advanced methods. Overall, the purpose of the research is to find ways to improve preservice teacher instruction on teaching science and engineering skills using self-regulated learning. If teacher educators can understand how preservice teachers perceive teaching methods of these practices and processes, teacher education can be designed in a way that is helpful for preservice teachers to be able to plan and implement the techniques successfully. Do preservice teachers understand that they do not need to teach all of the practices at the same time? Do preservice teachers understand that they can gradually introduce SRL to students by modeling and then using formal language and reflection?

Recall from prior research in Chapter 12 that preservice teachers may be more ready to learn about teaching science and engineering practices and SRL than previously thought. Recommendations from the research include building self-beliefs and task value for teaching science and engineering practices and SRL (Bråten & Strømsø, 2005; Tran et al., 2022). Case study allows for researchers to investigate these beliefs in detail and find out what preservice teachers offer as a rationale for their thinking. Additionally, it was previously found that preservice teachers were able to plan for most science and engineering practices and SRL, but needed support from teacher educators (Perry et al., 2008; Tran et al., 2022). Case study methods can help explicate why preservice teachers had difficulty asking students to create testable questions, revise their question and methods, participate in peer review, and disseminate their results to their peers or the larger scientific community. From the results, teacher educators can act to improve teacher education instruction in these areas.

The conceptual framework for this study plan is grounded in two frameworks that are hierarchal. Social cognitive theory (Bandura, 2002) serves as the superordinate framework that explains interactions between self-beliefs, behaviors, and environment. Pedagogical content knowledge (Schulman, 1986) serves as a subordinate framework that explains how teachers translate their own knowledge into a classroom learning environment for students. Social cognitive theory is useful as a foundation for the study because it can explain the internal variables and relationships such as self-beliefs and metacognition about science and engineering practice teaching and student SRL support, and external variables such as teacher behavior on planning for lessons with science and engineering practices supported by SRL and the interactions of learning in the methods class and lesson planning products. Pedagogical content knowledge is the intersection of a teacher's subject matter knowledge and their knowledge about teaching that subject matter (Schulman, 1986). Teachers would have informed pedagogical content knowledge if they were able to translate the subject knowledge they have to their design and ensure successful implementation of a learning environment for students who are non-experts in the subject matter. Framing pedagogical content knowledge for the study is needed to explain how teachers take what they learn in the methods class and convert it to lesson plans for a learning environment that supports science and engineering practices through SRL support.

The methods of the case study involve gathering multiple sources of information as data: interviews, teaching observations, and lesson plan artifact analysis. Interviews are considered to be self-report data, so observing lessons and gathering artifacts will help with trustworthiness. Since this study focuses on both science and engineering practices and SRL, the interview questions need to be constructed to ask about both topics, the observation protocol needs to note opportunities for both topics, and the codes for the transcripts and artifact analysis need to address both topics. The Next Generation Science Standards (NGSS Lead States, 2013) serves as a lens from which to identify instances of teacher interaction with science and engineering practices. In addition to having the eight science and engineering practices from the standards in mind, a researcher may want to drill down more deeply into the practices by using a task analysis tool. Perry et al. (2008) offer a helpful framework from which to construct the interviews, observations, and artifact analysis for the SRL factors being studied. They found ten intentional experiences that were built into university instruction and into field experiences (as described in Chapter 12). These ten experiences form a lens through which interview questions, observation protocol, and coding schemes can be developed from prior empirical work. Basing the data collection tools on prior research grounds the questions empirically, builds on ideas that already exist in the field, and also helps guard against validity threats.

Because researchers are human, there will always be some type of bias and reflexivity (influence of the researcher on the environment) introduced into any design, qualitative, quantitative, and mixed methods. However,

Research Designs for Science and Engineering Practices & SRL 247

Maxwell (2012) offers eight ways to guard against validity threats in qualitative research: (a) approach the research as a detective solving a crime, (b) search for negative cases and discrepant events, (c) triangulation, (d) gathering feedback from others, (e) member checks, (f) gathering detailed and complete data, (h) use descriptive statistics as an alternate lens for conjectures, and (i) compare multiple sites for the research. This study can use all of these approaches to examine ways in which conjectures from the data can be wrong or misleading.

For example, the researcher can approach the study as seeking clues for the ways that preservice teachers learn about and then use science and engineering practices, SRL, and the integration of the two in their instruction. The clues can be actualized by finding out in what ways the preservice teachers learn, how they transfer it into planning documents, and then how they implement it in a field experience. In searching for these clues, a researcher should keep an open mind for shadow knowledge – the ways that the preservice teachers *do not* learn, plan for, and teach aspects of the science and engineering practices through SRL. Triangulation is built into the methods by gathering multiple pieces of evidence from both self-report (internal) and observation (external) sources. The researcher can gather feedback by asking another researcher not associated with this project to check their coding and conjectures before moving on to coalescing their findings. Member checks with the inservice teachers can be conducted when the researcher interprets the data to be sure that the interpretation is aligned to the preservice teachers' understanding. Using prior empirical research from the fields of science and engineering practices and SRL is helpful in gathering detailed and complete data. However, the researcher may want to also continue gathering data until they achieve saturation, which is the point where the information gathered continues to repeat prior ideas. Quantifying findings can help a researcher see things that they may not see in the narrative, such as counting instances the preservice teachers interact with asking questions in the three situations of learning, planning, and teaching. Quantifying the findings can help summarize big picture ideas that may be hidden in the details of the narrative. Finally, conducting the study at different sites may help illuminate the findings that are dependent on context or available resources.

14.2 Quantitative Design Comparing Students in Two SRL Treatments

A research design that employs a quantitative comparison affords the opportunity to look across many students to find out how much they learn about science and engineering practices in two different conditions, one that integrates implicit SRL instruction and another that integrates explicit SRL instruction. By doing so, educational researchers can find out more about instructional approaches and refine effective instructional processes. Prior research has shown that explicit SRL instruction was helpful for learning SRL (Perry et al., 2008) and we know that explicit SRL instruction integrated into epistemic science knowledge is helpful for learning subject matter content (Peters & Kitsantas, 2010), but we do not yet know if explicit SRL instruction is helpful for learning all of the science and engineering skills, particularly in the middle school setting. Figure 14.3 displays the research design diagram for this study.

The conceptual framework for this study is Zimmerman's (2000) self-regulated learning theory and the framework of the science and engineering practices from NGSS. Zimmerman's SRL framework explicates the processes of SRL that are relevant to measure in the study. It also explains the various processes that occur during the forethought, performance, and self-reflection phases of learning. The processes that occur during these phases are intended to be measured with the SRL microanalysis method. Because this study focuses on middle school learning, the NGSS standards for grades 6–8 will be in the focus for the multidimensional measures used to assess what students know about science and engineering practices. The goals and conceptual frameworks help to form the research questions. RQ1: What are the outcomes of student learning of science and engineering practice instruction integrated with implicit SRL instruction and explicit SRL instruction? RQ2: What are the outcomes of student SRL knowledge when receiving either implicit SRL instruction and explicit SRL instruction?

The method of study is a two-group pre-and post-test comparison of student knowledge about science and engineering practices and SRL. Student selection for the study comes from multiple schools and teachers to reduce school effect and teacher effect. Student selection for the treatments should also be randomized as much as possible. Since student classes are typically already formed, then the randomization can happen at the class level rather than the student level.

Two treatments are the focus of the study. Treatment one will be implicit SRL instruction during science and engineering practice instruction and treatment two will be explicit SRL instruction during science and engineering practice instruction. Teachers will attend a professional development program for their prospective approach to teaching SRL and science and engineering practices to ensure fidelity with the intended

Research Designs for Science and Engineering Practices & SRL

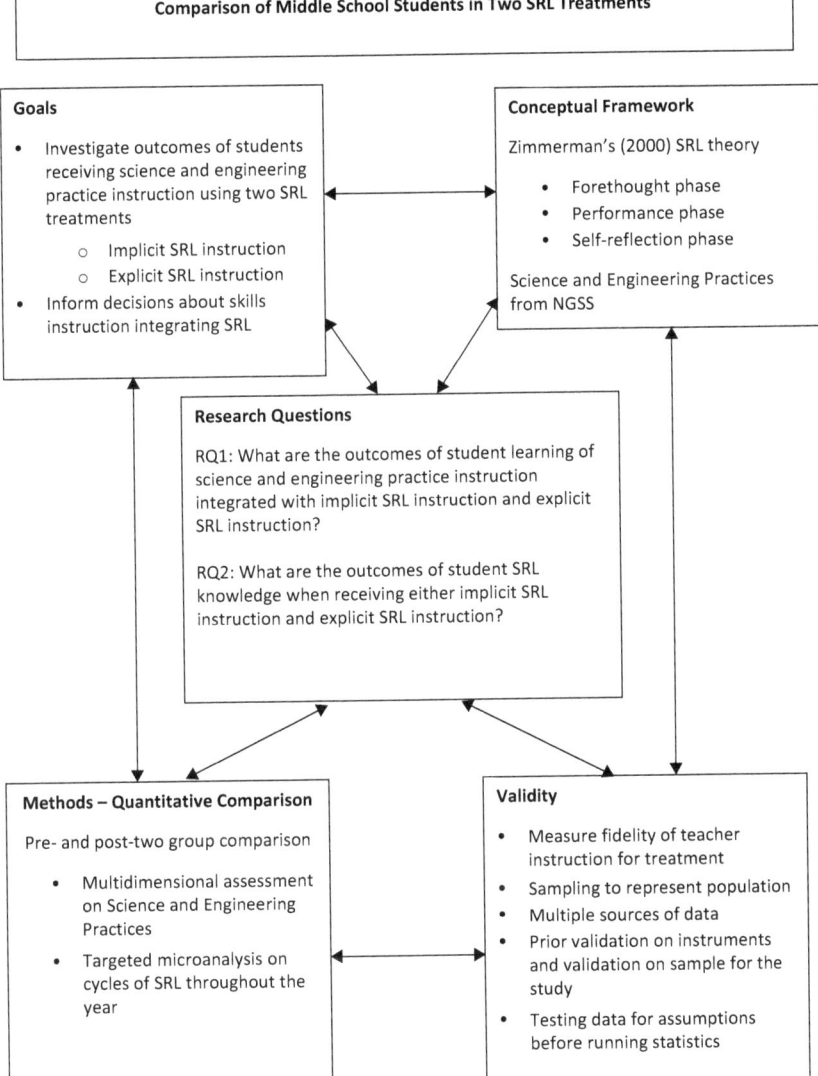

Figure 14.3   Research design diagram for quantitative comparison study

treatments. NGSS recommends that science and engineering practices be evaluated using multiple assessments, and assessing skills with a paper and pencil type multiple-choice test is not an accurate reflection of the types of skills in science and engineering practices. Rather, employing a validated performance assessment or rubric of student behavior for science and engineering practices over different days at the beginning of the year and at the end of the year would be a better reflection of student skills than a standardized test. SRL skills can be assessed in the same way as the science and engineering skills, but by using the microanalysis (Cleary, 2011) method. SRL microanalysis is a way to measure SRL skills during the time they are being used, allowing for measurement in the context of learning and at the time of learning.

Validity concerns for this study lie in the validation of measures, the standardized implementation of the treatments, and the randomization of the sample selection. If the measures are not valid and have some bias in a particular direction, then it is difficult for the groups to be compared. The same issue arises for the implementation of the treatments. If the teachers' implementation of the treatment is not consistent with the expectations for each treatment, then the comparison would be rendered inconclusive. Similarly, if the sample selection does not meet the assumptions needed for the statistical analysis techniques, then the results would be inconclusive. Figure 14.3 displays the research design diagram for this study.

14.3 Parallel Mixed Methods Design to Understand Inservice Teacher Learning of SRL for Science and Engineering Practices in a Professional Development Program

Mixed methods design of educational research provides the benefit of triangulating different kinds of data so that interpretation of results comes from different mindsets and perspectives (Teddlie & Tashakkori, 2009). Quantitatively based research methods allow for broad conjectures across many people and qualitatively based research methods offer deep analysis into a particular situation. When integrated together in a mindful way, qualitative and quantitative methods can lead to interpretation of findings that are greater than the sum of its parts. Figure 14.4 displays a research design diagram for a parallel mixed methods study of inservice teachers who are learning about integrating SRL into science and engineering practice instruction.

The goals of this study are formed from prior work in teacher education of science and engineering practices and SRL. There is literature that teachers find value in teaching both science and engineering practices and in SRL. However, the research on SRL shows that although teachers value their students understanding processes of SRL, they are unaware of

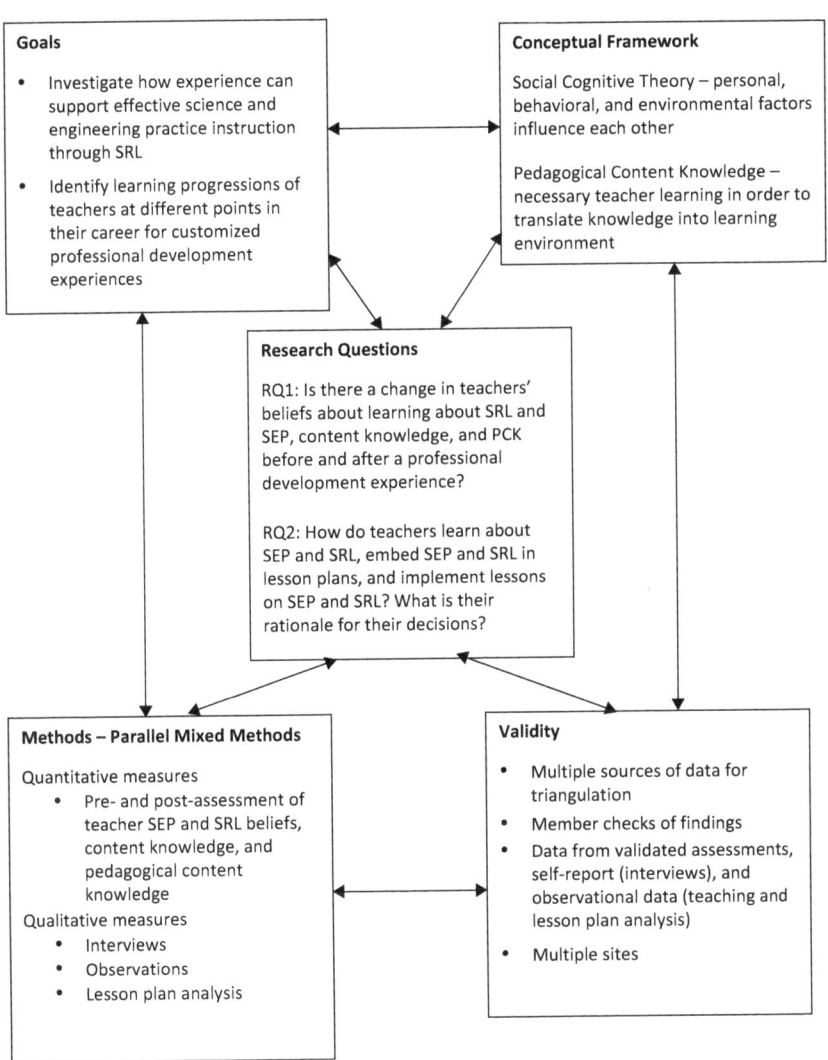

Figure 14.4 Research design diagram for mixed methods study

how to teach it and feel they do not have the time to teach SRL in their classes (Spruce & Bol, 2015). The research on science and engineering practices with inservice teachers shows that teachers embrace teaching the practices of developing and using procedures, analyzing and interpreting data, constructing explanations, using arguments with evidence, and evaluating and communicating information. However, there tends to be less instruction on asking questions and defining problems, using conceptual models, using mathematical and computational thinking, even when attending a professional development course focused on the practice. The prior research shows that there is uneven learning across the practices and across SRL processes, so there is still work to be done.

The conceptual framework for this study plan is grounded in two frameworks that are hierarchal. Social cognitive theory (Bandura, 2002) would serve as the superordinate framework that explains interactions between self-beliefs, behaviors, and environment. Pedagogical content knowledge (Schulman, 1986) would be a subordinate framework that could explain how teachers translate their own knowledge into a classroom learning environment for students. Social cognitive theory serves as a solid foundation for the study because it can explain the internal variables and relationships such as self-beliefs and metacognition about science and engineering practice teaching and student SRL support, and external variables such as teacher behavior for planning lessons with science and engineering practices supported by SRL and the interactions of learning in the professional development and implementation in classrooms. Pedagogical content knowledge is the intersection of a teacher's subject matter knowledge and their knowledge about teaching that subject matter (Schulman, 1986). Framing pedagogical content knowledge for the study can explain how teachers take what they learned in the professional development course and convert it to lesson plans for a learning environment that supports science and engineering practices through SRL support. The goals and the conceptual framework lay the foundation for the following research questions for the study:

RQ1: Is there a change in teachers' beliefs about learning about SRL and science and engineering practices, content knowledge, and pedagogical content knowledge before and after a professional development experience?
RQ2: How do teachers learn about science and engineering practices and SRL, embed science and engineering practices and SRL in lesson plans, and implement lessons on science and engineering practices and SRL? What is their rationale for their decisions?

The methods for this study are a parallel mixed method approach, where the quantitative data collection occurs concurrently with the qualitative data collection (Teddlie & Tashakkori, 2009). The reason for concurrent collection of data is to find information sources that may reveal different aspects of teachers' learning and application of science and engineering practices and SRL. The data sources are intended to inform each other. The qualitative data can be quantified and compared to quantitative results, while the quantitative data can be qualified and compared to qualitative results. This may reveal some discrepant events or reinforce trends that will elaborate the findings. The intention of the mixed methods design is not to take each type of data source and analyze separately then compare, but to have integration from data collection to analysis to findings. Figure 14.4 displays the research design diagram for this study.

The quantitative data sources for this study are a validated assessment of teacher beliefs about teaching science and engineering practices and SRL, subject matter knowledge of science and engineering practices and SRL, and pedagogical content knowledge of science and engineering practices and SRL. These measures help identify any differences in teachers before the professional development experience, after the professional development experience, and after implementation of the intended instructional approaches. The qualitative data sources would be interviews before the professional development experience, after the professional development experience, and after implementation of the intended teaching approaches. Additionally, lesson plan artifact analysis and classroom observations would be conducted to measure the quality of instruction to determine the level of implementation of the professional development experience in the teachers' classroom.

Maxwell's (2012) approaches to test conjectures from the data for accuracy can be employed in this study. For example, the researcher can approach the study as seeking clues for the ways that preservice teachers learn about and then use science and engineering practices, SRL, and the integration of the two in their instruction. The clues can be actualized by finding out in what ways the preservice teachers learn, how they transfer it into planning documents, and then how they implement it in a field experience. In searching for these clues, a researcher should keep an open mind for shadow knowledge – the ways that the teachers *do not* learn, plan for, and teach aspects of the science and engineering practices through SRL. Triangulation is built into the methods by gathering multiple pieces of evidence from both self-report (internal) and observation (external) sources. The researcher can gather feedback by asking another researcher not associated with this project to check their coding and conjectures before moving on to coalescing their findings. Member checks with the inservice teachers can be conducted when the researcher interprets the data to be sure that the interpretation is aligned to the preservice teachers' understanding. Qualifying the quantitative findings and quantifying the qualitative findings can help a researcher find instances of discrepant events or can emphasize themes that were hidden in the individual data sets. Finally, conducting the study at different sites may help illuminate the findings that are dependent on context or available resources.

14.4 Concluding Remarks about Supporting Science and Engineering Students with Self-Regulated Learning

There is evidence that SRL can be very helpful in learning a variety of knowledge and skills in diverse contexts. In addition, understanding science and engineering practices can help people solve problems in a systemic and rigorous way. Information and misinformation are at our fingertips at all times. Funds of knowledge no longer lie in what a person knows but in how they can find reliable information, digest it, and communicate it. Building student skills in science and engineering practices and building an awareness of SRL processes is an important part of helping students to become lifelong learners and problem solvers.

We should take every opportunity to teach preservice teachers, inservice teachers, and students how to recognize and act on their own learning processes and how science and engineering practices are conducted to create valid and reliable knowledge. When science and engineering practices and SRL are used in conjunction, there is an opportunity for students who do not yet feel part of the science and engineering community to join as full participants. Our aims as researchers and educators should be to reach all students so that they are proficient in being able to explore their world as they see fit. It is through teacher education and educational research that we can refine these ideas and continue to improve science and engineering education for all. Thank you for taking this journey with me.

References

Accreditation Board for Engineering and Technology (ABET). (2021). *About ABET*. www.abet.org/about-abet/

Akerson, V. L., Buzzelli, C. A., & Donnelly, L. A. (2017). Early childhood teachers' view of nature of science: The influence of intellectual levels, cultural values, and explicit reflective teaching. *Journal of Research in Science Teaching, 45*(6), 748–770.

Ames, C. (1992). Achievement goals and the classroom motivational climate. In D. H. Schunk & J. L. Meece (Eds.), *Student perceptions in the classroom* (pp. 327–348). Lawrence Erlbaum.

Augustine, N. R. (2005). *Rising above the gathering storm: Energizing and employing America for a brighter economic future*. National Academies Press, Washington, DC.

Bandura, A. (2002). Social cognitive theory in cultural context. *Applied Psychology, 51*(2), 269–290.

Brand, B. R. (2020). Integrating science and engineering practices: Outcomes from a collaborative professional development. *International Journal of STEM Education, 7*(13), 1–13. https://doi.org/10.1186/s40594-020-00210-x

Bråten, I., & Strømsø, H. I. (2005). The relationship between epistemological beliefs, implicit theories of intelligence, and self-regulated learning among Norwegian postsecondary students. *British Journal of Educational Psychology, 75*(4), 539–565.

Bybee, R. W., Taylor, J. A., Gardner, A. et al. (2006). *The BSCS 5E instructional model: Origins and effectiveness*. Office of Science Education, National Institutes of Health.

Christian, K. B., Kelly, A. M., & Buggalo, M. F. (2021). NGSS-based teacher professional development to implement engineering practices in STEM instruction. *International Journal of STEM Education, 8*(21), 1–18. https://doi.org/10.1186/s40594-021-00284-1

Cleary, T. J. (2011). Emergence of self-regulated learning microanalysis: Historical overview, essential features, and implications for research and practice. In D. H. Schunk & B. Zimmerman (Eds.), *Handbook of self-regulation of learning and performance* (2nd ed., pp. 329–345). Routledge/Taylor & Francis Group.

Cleary, T., Kitsantas, A., Peters-Burton, E. E. et al. (2022). Self-regulated learning professional development: Shifts and variations in teacher outcomes and approaches to implementation. *Teaching and Teacher Education, 111*, Article 103619. https://doi.org/10.1016/j.tate.2021.103619

Colclasure, B. C., Durham Brooks, T., Helikar, T., King, S. J., & Webb, A. (2022). The effects of a modeling and computational thinking professional development program on stem educators' perceptions toward teaching science and engineering practices. *Education Sciences, 12*(8), 570. https://doi.org/10.3390/educsci12080570

Collins, A., Brown, J. S., & Newman, S. E. (1989). Cognitive apprenticeship: Teaching the crafts of reading, writing, and mathematics. In L. B. Resnick (Ed.), *Knowing, learning, and instruction: Essays in honor of Robert Glaser* (pp. 453–494). Lawrence Erlbaum.

Corno, L., & Mandinach, E. (1983). The role of cognitive engagement in classroom learning and motivation. *Educational Psychologist, 18*(2), 88–108. http://dx.doi.org/10.1080/00461528309529266

Deci, E. L. (1975). *Intrinsic motivation*. Plenum Press.

Dembo, M. H. (2001). Learning to teach is not enough: Future teachers also need to learn how to learn. *Teacher Education Quarterly, 28*(4), 23–35.

DiBenedetto, M. K., & Zimmerman, B. J. (2010). Differences in self-regulatory processes among students studying science: A microanalytic investigation. *The International Journal of Educational and Psychological Assessment, 5*(1), 2–24.

Dignath van Ewijk, C., & van der Werf, M. P. C. (2012). What teachers think about self-regulated learning: Investigating teacher beliefs and teacher behavior of enhancing students' self-regulation. *Education Research International*, Article 741713. http://dx.doi.org/10.1155/2012/741713

Duschl, R. A., & Bybee, R. W. (2014). Planning and carrying out investigations: An entry to learning and to teacher professional development around NGSS science and engineering practices. *International Journal of STEM Education, 1*(12), 1–9. https://doi.org/10.1186/s40594-014-0012-6

Dweck, C. S., & Leggett, E. L. (1988). A social-cognitive approach to motivation and personality. *Psychological Review, 95*(2), 256–273. https://doi.org/10.1037/0033-295X.95.2.256

Efklides, A. (2011). Interactions of metacognition with motivation and affect in self-regulated learning: The MASRL model. *Educational Psychologist, 46*(1), 6–25. https://doi.org/10.1080/00461520.2011.538645

Feldon, D. F., Timmerman, B. C., Stowe, K. A., & Showman, R. (2010). Translating expertise into effective instruction: The impacts of cognitive task analysis (CTA) on lab report quality and student retention in the biological sciences. *Journal of Research in Science Teaching, 47*(10), 1165–1185.

Flores, I. M. (2015). Developing preservice teachers' self-efficacy through field-based science teaching practice with elementary students. *Research in Higher Education Journal, 27*, 1–19.

Foster, I. (2006). 2020 computing: A two-way street to science's future. *Nature*, *440*(7083), 419.

French, D. A., & Burrows, A. C. (2018). Evidence of science and engineering practices in preservice secondary science teachers' instructional planning. *Journal of Science Education and Technology*, *27*, 536–549. https://doi.org/10.1007/s10956-018-9742-4

Fuller, F. F., & Bown, O. H. (1975). Becoming a teacher. In K. Ryan (Ed.), *Teacher education: The 74th yearbook of the national society for the study of education* (pp. 25–52). University of Chicago Press.

Gordon, S. C., Dembo, M. H., & Hocevar, D. (2007). Do teachers' own learning behaviors influence their classroom goal orientation and control ideology? *Teaching and Teacher Education*, *23*(1), 36–46.

Guzey, S. S., Roehrig, G., Tank, K., Moore, T., & Wang, H. H. (2014). A high-quality professional development for teachers of grades 3–6 for implementing engineering into classroom. *School Science and Mathematics*, *114*(3), 139–149.

Hang, N. T. T., & Srisawasdi, N. (2021). Perception of the Next Generation Science Standard instructional practices among Vietnamese pre-service and in-service teachers. *Journal of Technology and Science Education*, *11*(2), 440–456. https://doi.org/10.3926/jotse.1154

Hogan, K. (2000). Exploring a process view of students' knowledge about the nature of science. *Science Education*, *84*(1), 51–70.

Khishfe, R., & Abd-El-Khalick, F. (2002). The influence of explicit reflective versus implicit inquiry-oriented instruction on sixth graders' views of nature of science. *Journal of Research in Science Teaching*, *39*(7), 551–578.

Klusmann, U., Kunter, M., Trautwein, U., Ludtke, O., & Baumert, J. (2008). Engagement and emotional exhaustion in teachers: Does the school context make a difference? *Applied Psychology*, *57*(s1), 127–151. https://doi.org/10.1111/j.1464-0597.2008.00358.x

Kramarski, B. (2008). Promoting teachers' algebraic reasoning and self-regulation with metacognitive guidance. *Metacognition and Learning*, *3*(2), 83–99.

Kramarski, B., & Kohen, Z. (2017). Promoting preservice teachers' dual self-regulation roles as learners and as teachers: Effects of generic vs. specific prompts. *Metacognition & Learning*, *12*(2), 157–191. https://doi.org/10.1007/s11409-016-9164-8

Kramarski, B., & Michalsky, T. (2009). Investigating preservice teachers' professional growth in self-regulated learning environments. *Journal of Educational Psychology*, *101*(1), 161–175. https://doi.org/10.1037/a0013101

(2010). Preparing preservice teachers for self-regulated learning in the context of technological pedagogical content knowledge. *Learning and Instruction*, *20*(5), 434–447.

Kramarski, B., & Revach, T. (2009). The challenge of self-regulated learning in mathematics teachers' professional training. *Educational Studies in Mathematics*, *72*(3), 379–399.

Kremer-Hayon, L., & Tillema, H. H. (1999). Self-regulated learning in the context of teacher education. *Teaching and Teacher Education, 15*(5), 507–522.

Loucks-Horsley, S., Love, N., Stiles, K. E., Mundry, S. E., & Hewson, P. (2003). *Designing professional development for teachers of science and mathematics* (2nd ed.). Corwin Press.

Lucangeli, D., & Cabrele, S. (2006). Mathematical difficulties and ADHD. *Exceptionality, 14*(1), 53–62. https://doi.org/10.1207/s15327035ex1401_5

Maxwell, J. A. (2012). *Qualitative research design: An interactive approach.* Sage Publications.

McComas, W. (2019). Principal elements of nature of science: Informing science teaching while dispelling the myths. In W. F. McComas (Ed.), *The nature of science in science instruction: Rationales & strategies* (pp. 35–66). Springer.

Merritt, E., Chiu, J. L., Peters-Burton, E. E., & Bell, R. (2018). Teachers' integration of scientific and engineering practices in primary classrooms. *Research in Science Education, 48*(6), 1321–1337. https://doi.org/10.1007/s11165-016-9604-0

Moos, D. C., & Ringdal, A. (2012). Self-regulated learning in the classroom: A literature review on the teacher's role. *Education Research International, 2012*, 423284. https://doi.org/10.1155/2012/423284

National Academy of Engineering & National Research Council (NAE & NRC). (2009). *Engineering in K-12 education: Understanding the status and improving the prospects.* National Academies Press.

National Research Council. (1996). *National science education standards.* National Academies Press.

NGSS Lead States. (2013). *Next generation science standards: For states, by states.* The National Academies Press.

NGSS@NSTA (2014). *Science and engineering practices.* Retrieved from https://ngss.nsta.org/PracticesFull.aspx.

Osborne, J. (2014). Scientific practices and inquiry in the science classroom. In N. G. Lederman & S. K. Abell (Eds.), *Handbook of research on science education* (vol. 2, pp. 579–599). Routledge.

Osborne, J., Collins, S., Ratcliffe, M., Millar, R., & Duschl, R. (2003). What "ideas-about-science" should be taught in school science? A Delphi study of the expert community. *Journal of Research in Science Teaching, 40*(7), 692–720. https://doi.org/10.1002/tea.10105

Pazhoman, H., & Sarkhosh, M. (2019). The relationship between Iranian English high school teachers' reflective practices, their self-regulation and teaching experience. *International Journal of Instruction, 12*(1), 995–1010.

Perry, N. E., Hutchinson, L., & Thauberger, C. (2008). Talking about teaching self-regulated learning: Scaffolding student teachers' development and use of practices that promote self-regulated learning. *International Journal of Educational Research, 47*(2), 97–108.

Perry, N. E., Phillips, L., & Hutchinson, L. (2006). Mentoring student teachers to support self- regulated learning. *The Elementary School Journal, 106*(3), 237–254.

Peters, E. E. (2009). *Thinking like scientists: Using metacognitive prompts to develop nature of science knowledge.* Verlag Dr. Müller Aktiengesellschaft & Co. KG Publishers.

(2012). Developing content knowledge in students through explicit teaching of the nature of science: Influences of goal setting and self-monitoring. *Science & Education, 21*(6), 881–898. https://doi.org/10.1007/s11191-009-9219-1

Peters, E. E., & Kitsantas, A. (2010). Self-regulation of student epistemic thinking in science: The role of metacognitive prompts. *Educational Psychology, 30*(1), 27–52. https://doi.org/10.1080/01443410903353294

Peters-Burton, E. E. (2015). Outcomes of a self-regulatory curriculum model: Network analysis of middle school students' views of nature of science. *Science & Education, 24*(7–8), 855–885. https://doi.org/10.1007/s11191-015-9769-3

(2017). Strategies for learning nature of science knowledge: A perspective from educational psychology. In M. R. Matthews (Ed.), *History, philosophy and science teaching: New perspectives* (pp. 167–193). Springer.

Peters-Burton, E. E., & Botov, I. S. (2017). Self-regulated learning microanalysis as a tool to inform professional development delivery in real-time. *Metacognition and Learning, 12*(1), 45–78. https://doi.org/10.1007/s11409-016-9160-z

Peters-Burton, E. E., & Burton, S. R. (2020). The use of metacognitive prompts to foster nature of science learning. In W. McComas (Ed.), *Nature of science in science instruction* (pp. 179–197). Springer. https://link.springer.com/chapter/10.1007%2F978-3-030-57239-6_9

Peters-Burton, E. E., Goffena, J., & Stehle, S. M. (2022). Utility of a self-regulated learning microanalysis for assessing teacher learning during professional development. *Journal of Experimental Education, 90*(3), 523–549. https://doi.org/10.1080/00220973.2020.1799314

Peters-Burton, E. E., Rich, P., Cleary, T. et al. (2020). Using computational thinking for data practices in high school science. *The Science Teacher, 87*(6), 30–36.

Peters-Burton, E. E., Rich, P. J., Kitsantas, A., Laclede, L., & Stehle, S. M. (2022). High school science teacher use of planning tools to integrate computational thinking. *Journal of Science Teacher Education, 33*(6), 598–620. https://doi.org/10.1080/1046560X.2021.1970088

Porter, A., & Peters-Burton, E. E. (2021). Investigating teacher development of self-regulated learning strategies for secondary science students. *Teaching and Teacher Education, 105*, 103403. https://doi.org/10.1016/j.tate.2021.103403

Pleasants, J., & Olson, J. K. (2019). What is engineering? Elaborating the nature of engineering for K-12 education. *Science Education, 103*(1), 145–166. https://doi.org/10.1002/sce.21483

Rohrkemper, M. (1989). Self-regulated learning and academic achievement: A Vygotskian view. In B. J. Zimmerman & D. H. Schunk (Eds.), *Self-regulated learning and academic achievement: Theory, research and practice* (pp. 143–167). Springer.

Ryan, R. M., Connell, J. P., & Deci, E. L. (1984). A motivational analysis of self-determination and self-regulation in education. In C. Ames & R. Ames (Eds.), *Research on motivation in education* (Vol. 2, pp. 13–52). Academic Press.

Schulman, L. S. (1986). Those who understand: Knowledge growth in teaching. *Educational Researcher, 15*(2), 4–14.

Seo, D., & Taherbhai, H. (2009). Motivational beliefs and cognitive processes in mathematics achievement, analyzed in the context of cultural differences: A Korean elementary school example. *Asia Pacific Education Review, 10*(2), 193–203. https://doi.org/10.1007/s12564-009-9017-0

Spruce, R., & Bol, L. (2015). Teacher beliefs, knowledge and practice of self-regulated learning. *Metacognition and Learning, 10*(2), 245–277. https://doi.org/10.1007/s11409-014-9124-0

Spuck, T. (2014). Putting the Authenticity in science learning. In T. Spuck, L. Jenkins, & R. Dou (Eds.), *Einstein fellows: Best practices in STEM education* (pp. 118–156). Peter Lang.

Stehle, S. M., & Peters-Burton, E. E. (2019). Developing student 21st century skills in selected exemplary inclusive STEM high schools. *International Journal of STEM Education, 6*(39), 1–15. https://doi.org/10.1186/s40594-019-0192-1

Teddlie, C., & Tashakkori, A. (2009). *Foundational of mixed methods research: Integrating quantitative and qualitative approaches in the social and behavioral sciences.* Sage Publications.

Tran, H. H., Capps, D. K., & Hodges, G. W. (2022). Preservice science teachers' perspectives on and practices related to self-regulated learning after a brief learning opportunity. *Sustainability, 14*(10), 5923. https://doi.org/10.3390/su14105923

Tricarico, K., & Yendol-Hoppey, D. (2012). Teacher learning through self-regulation: An exploratory study of alternatively prepared teachers' ability to plan differentiated instruction in an urban elementary school. *Teacher Education Quarterly, 39*(1), 139–158.

Wang, M. C., & Peverly, S. T. (1986). The self-instructive process in classroom learning contexts. *Contemporary Educational Psychology, 11*(4), 370–404. http://dx.doi.org/10.1016/0361-476X(86)90031-7

Weintrop, D., Beheshti, E., Horn, M. et al. (2016). Defining computational thinking for mathematics and science classrooms. *Journal of Science Education and Technology, 25*(1), 127–147. https://doi.org/10.1007/s10956-015-9581-5

Wigfield, A., & Eccles, J. S. (1992). The development of achievement task values: A theoretical analysis. *Developmental Review, 12*(3), 265–310. https://doi.org/10.1016/0273-2297(92)90011-P

Wing, J. M. (2006). Computational thinking. *Communications of the ACM, 49*(3), 33–35. https://doi.org/10.1145/1118178.1118215

Yin, R. K. (2003). *Case study research: Design and methods.* Sage Publications.

Yong, B. (2014, May 7). *Man-made electromagnetic noise disrupts a bird's compass*. National Geographic. Man-Made Electromagnetic Noise Disrupts a Bird's Compass (nationalgeographic.com)

Zimmerman, B. J. (2000). Attaining self-regulation: A social-cognitive perspective. In M. Boekaerts, P. Pintrich, & M. Zeidner (Eds.), *Handbook of self-regulation* (pp. 13–39). Academic Press.

(2008). Investigating self-regulation and motivation: Historical background, methodological developments, and future prospects. *American Educational Research Journal*, 45(1), 166–183.

Zimmerman, B. J., & Kitsantas, A. (1997). Developmental phases in self-regulation: Shifting from process to outcome goals. *Journal of Educational Psychology*, 89(1), 29–36.

(2002). Acquiring writing revision and self-regulatory skill through observation and emulation. *Journal of Educational Psychology*, 94(4), 660–668. http://dx.doi.org/10.1037/0022-0663.94.4.660

(2014). Comparing the predictive power of self-discipline and self-regulation measures of learning. *Contemporary Educational Psychology*, 39(2), 145–155. http://dx.doi.org/10.1016/j.cedpsych.2014.03.004

Index

5E lesson planning format, 11
5E Model of Curriculum Design, 228

abstraction, 15
academic risks, 31
academic strategies, 36
adapt, 11, 33, 35, 41, 53, 66, 74, 98, 107, 108, 120, 153, 174, 195, 218, 233
addresses the problem, 21, 164
algorithmic thinking, 15, 89, 132, 142
analyze data, 9, 66, 118, 131, 210
analyze the practice, 11
Anomalous data, 19
argumentation, 14, 15, 164, 209, 214, 223, 227, 231, 232
ask questions, 9, 14, 19, 26, 34, 47, 49, 51, 53, 55, 65, 66, 106, 185, 229
asking questions, 4, 11, 13, 16, 18, 26, 47, 48, 49, 50, 51, 52, 55, 56, 57, 58, 59, 60, 61, 62, 64, 68, 87, 89, 106, 108, 127, 129, 142, 160, 162, 176, 182, 183, 203, 214, 229, 232, 236, 247, 252
attainable goal, 38
attribution, 33, 35, 41, 75, 84, 85, 140, 232
attributions, 33, 37, 41, 86, 133, 233
automated, 26, 33, 117
automation, 15
avoids, 35

background knowledge, 5, 6, 7, 48, 49, 66
become more aware, 28, 40, 129, 205
beliefs, 19, 31, 36, 56, 77, 100, 102, 123, 155, 177, 197, 198, 210, 212, 213, 214, 216, 245, 252, 253, 255, 256, 260
bias, 16, 52, 191, 241, 246, 250
black box, 29
breakdown, 39

cardboard, 62, 63, 204
career, 3, 4, 28, 37, 85, 129, 161, 215, 216

case studies, 47, 61, 66, 83, 88, 105, 108, 110, 126, 131, 139, 143, 159, 162, 164, 180, 183, 185, 201, 205
case study, 243
cataloged, 36
clarify, 47, 51, 53, 229
classroom, 11, 32, 37, 38, 39, 48, 58, 62, 64, 66, 78, 107, 108, 129, 132, 136, 142, 162, 166, 183, 205, 210, 212, 216, 217, 245, 252, 253, 255, 256, 257, 258, 260
coaching, 42, 47, 54, 55, 56, 57, 59, 60, 66, 75, 77, 78, 80, 81, 88, 98, 100, 102, 103, 107, 108, 120, 121, 122, 123, 124, 134, 136, 137, 143, 152, 155, 157, 164, 175, 176, 177, 179, 185, 196, 197, 198, 200
coding qualitative data, 25
collaborative, 5, 18, 53, 69, 70, 71, 73, 75, 80, 88, 90, 91, 97, 154, 177, 186
communicate results, 9
communicating, 4, 6, 7, 13, 21, 101, 185, 186, 187, 189, 190, 192, 195, 196, 197, 198, 199, 200, 201, 203, 204, 205, 206, 210, 215, 230, 231, 236, 252
computational thinking, 6, 7, 14, 15, 19, 21, 25, 26, 37, 131, 132, 133, 134, 136, 137, 138, 139, 140, 142, 185, 215, 230, 231, 252, 256, 259
conceptual models, 16
confusion, 63
Congruence of goals, 38
constraints, 19, 21, 47, 48, 50, 52, 53, 58, 60, 61, 62, 64, 143, 145, 146, 164, 168, 169, 203
consumers, 15, 21
content knowledge, 3, 5, 8, 10, 22, 35, 42, 217, 219, 223, 245, 252, 253, 257, 259
contexts, 3, 7, 14, 24, 27, 160, 254, 260
controllable sources of attribution, 41
create explanations, 15
creating scales, 25

creative, 16, 212
creativity, 5, 18, 19
critique, 21, 170, 171, 172, 216
curiosity, 5, 18, 19
curious, 16, 105
cycle, 10, 22, 32, 33, 34, 36, 37, 47, 48, 54, 66, 67, 75, 88, 89, 97, 110, 111, 115, 119, 131, 134, 143, 144, 145, 148, 152, 164, 165, 174, 185, 186, 195, 223, 229
cyclical design process, 6
cyclical phases of SRL, 10

decompose, 24, 63, 141, 170, 172, 233
decomposition, 15, 25, 132, 133, 136, 137, 138, 139, 140, 141, 142, 230
defensive reaction, 40
define problems, 14, 49, 50, 55, 57, 65, 66
describe models, 15
describing a phenomenon, 14
design action research project, 12
design for an electromagnet, 154
design solutions, 21, 68, 70, 90, 92, 113, 114, 118, 122, 130, 131, 143, 146, 150, 153, 159, 163, 185
develop deep knowledge, 16
develop models, 20
developing models, 18
developing procedures, 4
developmentally appropriate, 15, 58, 161
disadvantage, 13
disciplinary approaches, 3, 4
disciplinary aspects, 16
disciplinary characteristics, 13
disciplinary thinking, 8, 19

ecologically friendly, 62
educational psychology, 3, 22, 259
educational research, 5, 14, 16, 29, 241, 250, 254
educational research frameworks, 5
educational researchers, 12
efficient, 3, 32, 43, 125, 132, 143, 199, 217, 223, 227
elaborates, 35, 110, 131
elaboration, 232
electrolysis, 78
electromagnet, 26, 154, 155, 156, 157, 158, 160, 161
elegance, 21
empirical data, 15
empirical evidence, 5, 14, 19, 51, 52
empower students, 28
empowering, 26
emulate, 42

emulation, 42, 55, 58, 60, 76, 77, 78, 80, 84, 98, 102, 108, 120, 123, 135, 136, 138, 153, 155, 160, 161, 175, 177, 196, 198, 261
engagement, 229
engineering design processes, 22
engineering educators, 6
engineering practices, 19
engineering teachers, 19
entry point to learning, 55, 75, 98, 120, 153, 174, 195
epistemic knowledge, 8, 10, 16, 22, 31, 43, 48, 59, 67, 81, 89, 102, 111, 124, 144, 156, 165, 178, 186, 199
epistemologies, 35
ethical interpretation of the evidence, 19
ethical standards, 5, 18
evaluate the outcomes, 21
evaluation, 40, 53, 54, 61, 62, 63, 74, 83, 85, 95, 106, 118, 133, 139, 140, 149, 152, 159, 160, 161, 162, 169, 170, 172, 192, 194, 202, 203, 204, 205, 206, 213, 227, 229, 232, 235, 236, 237
examples, 36
experimental design investigations, 22
expert and novice learners, 32
explanation, 230
explicit, 36
exploration, 230

fade all support, 42
failing, 31
failure, 33, 35, 56, 96, 97, 101, 128, 141, 175
feedback, 24, 32, 34, 35, 38, 40, 42, 55, 77, 84, 85, 91, 106, 141, 160, 161, 162, 181, 182, 183, 202, 204, 205, 211, 212, 213, 218, 219, 231, 232, 247, 254
find patterns, 15, 115, 131
flexible thinker, 16
focus, 5, 7, 8, 10, 22, 32, 35, 48, 49, 54, 59, 61, 62, 66, 75, 81, 83, 84, 94, 97, 99, 103, 105, 106, 119, 121, 124, 126, 127, 133, 136, 137, 139, 140, 152, 156, 159, 160, 161, 173, 178, 180, 181, 182, 194, 196, 199, 202, 203, 214, 215, 217, 218, 219, 223, 227, 228, 248
Forethought Phase, 30
form a question, 26, 52, 72
future opportunities, 37

gather evidence, 18
gathering evidence, 4
generalized, 39
goal orientation, 31, 58, 99, 101, 104, 105, 106, 194, 196, 202, 203, 213, 230, 257

graph, 39, 114, 115, 122, 132, 136, 138, 139, 200, 201, 202, 205, 206, 238
gravitational force, 52

habits of mind, 6
healthy skepticism, 5, 16, 19
heating ice investigation, 139
height of tides, 121
hierarchical organization, 38
high self-efficacy, 31, 36, 56
high task value, 31, 37, 159, 209
history, 5, 16, 259
how scientists and engineering think, 16
how to think efficiently, 29

implications for society, 21
influences, 28, 36, 163, 238
inquiry methods, 31
inservice teachers, 11, 12, 214, 215, 223, 243, 247, 250, 252, 254
instructional design, 16
integrate, 25, 26, 28, 132, 133, 137, 192, 210, 213, 214, 215, 259
intention of teaching science and engineering practices, 16
internal culture of engineering, 19
investigation, 139

key features for each practice, 39
K-W-L chart, 31

learn something new, 36
learning environments, 4, 8, 12, 40, 257
learning outcomes, 10, 99, 197, 215
learning strategies, 34, 36, 41, 43, 55, 56, 76, 77, 88, 99, 120, 129, 135, 153, 175, 196, 228, 259
learning task, 10, 15, 30, 31, 32, 33, 36, 41, 48, 49, 61, 64, 67, 73, 77, 82, 87, 89, 100, 101, 104, 107, 108, 109, 111, 117, 121, 122, 125, 130, 135, 138, 141, 142, 144, 154, 158, 159, 161, 162, 163, 165, 180, 181, 184, 186, 190, 197, 201, 206, 209, 217
learning theory, ix, 3, 28, 30
level of confidence, 36
level of detail, 4, 25
life-long learners, 28
logical, 7, 18, 21, 113, 114, 132, 164
low self-efficacy, 31, 36, 56, 176
low task value, 37, 161

major aim of science, 4
make improvements, 29
make meaning from the data, 15
making a claim, 15, 166, 179, 181

making claims, 18, 180
manipulated, 10, 62, 80
mastery, 13, 24, 31, 41, 56, 60, 82, 93, 99, 101, 104, 105, 107, 108, 115, 125, 138, 148, 158, 176, 179, 196, 199, 201, 202, 213, 229
mechanism for learning, 29
memorization, 38
memorizing, 34
metacognitively aware, 30
miniature golf course, 100
miniature golf course challenge, 100
mistakes, 29, 101, 104, 105, 106, 107, 108, 116, 122, 162, 183, 188, 202
model, 4
model science and engineering practices, 13
modeling, 26, 32, 36, 39, 40, 41, 42, 49, 55, 56, 57, 58, 60, 64, 68, 76, 77, 78, 80, 82, 84, 85, 86, 90, 93, 98, 99, 100, 101, 102, 104, 107, 108, 113, 115, 118, 120, 121, 122, 123, 125, 129, 134, 135, 136, 142, 147, 151, 153, 154, 155, 158, 160, 162, 166, 168, 171, 172, 175, 176, 177, 179, 183, 184, 190, 196, 198, 200, 205, 210, 212, 213, 214, 245, 256
monitor progress, 4, 7, 24, 29, 30, 32, 34, 39, 40, 55, 59, 70, 76, 81, 84, 99, 103, 120, 124, 127, 128, 129, 135, 137, 153, 156, 175, 178, 196, 199, 218, 223
MPI-S, 47, 54, 55, 56, 59, 60, 75, 77, 81, 82, 98, 99, 102, 103, 104, 120, 123, 124, 125, 134, 135, 137, 138, 152, 153, 156, 158, 174, 175, 176, 178, 179, 195, 196, 199, 201
murky ideas, 28

naïve learner, 34
natural world, 4, 7, 14, 18, 20, 47, 48, 51, 64, 66, 88, 143, 159, 161, 166
nature of engineering, 6
nature of science, 5, 7, 16, 19, 22, 35, 39, 48, 89, 111, 144, 165, 186, 219, 255, 257, 258, 259
necessity and sufficiency, 21
NGSS, 6, 8, 14, 15, 47, 48, 50, 66, 68, 88, 90, 110, 112, 131, 143, 145, 164, 166, 185, 186, 187, 214, 246, 248, 249, 255, 256, 258

observation, 42, 91, 113, 246, 247, 253, 261
opportunities, 25, 35, 39, 48, 56, 64, 67, 77, 86, 89, 99, 109, 111, 121, 128, 129, 135, 142, 144, 153, 162, 165, 175, 184, 186, 196, 203, 205, 210, 212, 213, 215, 218, 230, 246
optimal, 40, 43, 100, 101, 103, 105, 154

organize, 19
outcome and process goals, 22
outcome goal, 37
outcome goals, 22, 26, 38, 39, 40, 48, 68, 89, 111, 144, 165, 218, 261
ownership, 29, 35

packaging, 47, 57, 58, 62, 64
parallel circuit, 10
Parallel Mixed Methods Design, 250
parsimony, 21
pattern finding, 15
peer review, 5, 19, 73, 92, 95, 97, 103, 104, 106, 115, 118, 122, 124, 126, 179, 192, 193, 210, 218, 245
pendulum, 136
Performance Phase, 31
periodic motion of a pendulum, 100
persisted, 31, 219
phenomena, 4
phenomenon, 14, 16, 18, 19, 26, 48, 49, 50, 52, 53, 64, 67, 68, 69, 70, 71, 72, 73, 74, 75, 78, 79, 86, 88, 94, 112, 113, 114, 115, 116, 122, 124, 133, 145, 150, 160, 164, 166, 167, 187, 188, 191, 192, 193
physics, 28, 38, 101, 228
picture book, 177
potato chip packaging, 57, 58, 60, 61, 62, 63
preservice teachers, 4, 12, 209, 210, 211, 212, 213, 243, 244, 245, 247, 253, 254, 256, 257
prior expectations, 18
problem-solving, 4, 7, 10, 215
procedural knowledge, 8, 16, 22, 41, 43, 59, 81, 102, 124, 156, 178, 199
process goals, 22, 24, 26, 31, 32, 35, 37, 38, 40, 41, 47, 48, 50, 51, 53, 57, 58, 63, 64, 66, 67, 72, 74, 78, 80, 84, 85, 89, 92, 94, 95, 100, 102, 110, 111, 112, 113, 115, 118, 122, 127, 128, 131, 144, 145, 146, 148, 155, 162, 164, 165, 168, 170, 172, 177, 183, 186, 187, 189, 191, 193, 197, 198, 204, 218, 223
productive, 4, 30, 31, 32, 33, 34, 36, 37, 38, 41, 48, 61, 67, 89, 111, 144, 165, 186
professional development experiences, 12, 209, 216, 217
professional scientists, 5, 35
professional standards, 5, 18
prompts, 55, 59, 60, 75, 98, 120, 134, 152, 174, 195, 257, 259
prosper, 29
prototype, 21, 101, 103, 114, 179
proximal, 38, 48, 67, 89, 111, 144, 165, 186
purposeful, 16

qualities, 51
quantitative comparison, 247

rationale, 8, 49, 52, 57, 61, 77, 81, 83, 100, 104, 121, 126, 135, 138, 154, 158, 176, 180, 197, 201, 245, 252
reach the goal, 14, 38
reasoned, 36, 168
reasoning, 5, 15, 16, 53, 71, 72, 73, 95, 113, 114, 117, 148, 149, 154, 156, 164, 165, 167, 168, 169, 170, 171, 172, 173, 190, 192, 194, 257
recognizing constraints, 14
recording, 49, 91, 112, 132
recycling program, 197
refine models, 14
reflect on their outcomes, 30
reflective practitioner, 65, 87, 109, 130, 142, 163, 184, 206
reflects, 32
relationship between velocity and acceleration, 38
relationships, 20, 50, 51, 52, 57, 58, 62, 69, 70, 71, 72, 73, 74, 75, 78, 80, 81, 82, 83, 85, 86, 92, 115, 143, 146, 148, 149, 151, 154, 155, 156, 157, 159, 160, 163, 230, 231, 232, 246, 252
relevant observations, 28
research design framework, 241
research designs, 11
resilient, 31
resources, 14, 28, 37, 50, 108, 143, 190, 193, 194, 202, 203, 204, 215, 247, 254
respect evidence, 28
reviewing, 54, 95, 169

Science and engineering teachers, 12
Science and technology, 5
scientific claims, 5, 19, 131
scientific community, 5, 18, 143, 210, 245
scientific inquiry, 10, 34, 36
scientific knowledge production, 18
scientific questions, 6, 16, 34, 50, 51, 57, 112
scope of the problem, 14, 21
scoping a problem, 19
seeking help, 33
self-control, 42, 55, 59, 60, 76, 77, 81, 82, 99, 103, 104, 120, 124, 125, 135, 137, 138, 153, 156, 158, 160, 161, 175, 178, 180, 196, 199, 201
self-efficacy, 31, 36, 54, 56, 58, 59, 61, 62, 63, 77, 80, 86, 99, 121, 135, 140, 154, 173, 175, 176, 178, 181, 182, 183, 196, 209, 212, 215, 217, 218, 219, 230, 256
self-monitoring, 32, 37, 40, 41, 259

self-motivate, 4, 31
Self-Reflection Phase, 32
self-regulation, 33, 42, 55, 59, 76, 77, 81, 99, 103, 120, 124, 135, 137, 153, 157, 175, 179, 196, 200, 211, 255, 256, 257, 258, 260, 261
set goals, 14, 30, 34, 36, 38, 58, 80, 102, 121, 123, 127, 134, 135, 137, 154, 155, 177, 198, 218, 229
set productive goals, 4
setting goals, 37
share, 5, 18, 29, 39, 68, 81, 103, 124, 198, 230
similarities, 43, 113, 114, 243
simple circuits, 10
skillful learner, 35
SMART goals, 37
snack packaging design challenge, 47
solve problems, 15, 19, 21, 28, 35, 186, 233, 254
solving human needs, 7
sophisticated forethought, 34
specific goal, 37
specifications, 19, 21
spreadsheet, 37, 136
standard, 28, 32, 48, 64, 90, 112, 113
strategic plan, 31, 37, 126, 127, 128
student scientists, 5
student-centered, 26, 28, 212, 218, 228
suitable, 53
superficial level, 35
support human needs, 4

tailored to fit, 14
taking risks, 63, 99, 141
task analysis table, 25, 26
Task analysis tables, 25
taxonomy, 24, 43
teachable moment, 19
teacher education literature, 12
teacher educators, 12

teaching, explicitly and reflectively, 16
technology, 18
testable, 34, 48, 53, 57, 61, 64, 162, 210, 245
theories, 5, 18, 38, 148, 150, 241, 255
think like a scientist, 8, 28, 67
transfer, 3, 27, 211, 243, 247, 253
trials, 35, 39, 92, 108, 109, 112, 113
trust, 36, 40, 56, 176

uncontrollable sources of attribution, 41
unexpected outcomes, 19
unique questions, 16
units, 39, 74, 75, 132, 156, 206
unproductive, 4, 41
unsuitable, 53
use mathematics, 15, 131, 132, 142
user manual, 11

valid and reliable data, 19
valid and reliable knowledge, 28, 254
valid procedure, 33
valid results, 35
validate, 35
validated, 38, 249, 253
variables, 3, 10, 20, 26, 39, 50, 51, 52, 57, 58, 60, 61, 62, 63, 64, 69, 70, 71, 72, 73, 74, 75, 78, 79, 80, 81, 82, 83, 85, 86, 89, 92, 93, 94, 95, 96, 97, 100, 102, 103, 104, 105, 106, 109, 115, 116, 117, 123, 124, 132, 136, 137, 143, 146, 148, 149, 151, 154, 155, 156, 157, 159, 160, 162, 163, 230, 231, 232, 235, 238, 246, 252
variance, 39, 117
volleyball, 33

working in groups, 6, 204

zone of development, 38

Printed in the USA
CPSIA information can be obtained
at www.ICGtesting.com
LVHW021127131223
766288LV00005B/234

9 781009 108270